John Croumbie Brown

Forests and Forestry in Poland, Lithuania, the Ukraine, and the

Baltic Provinces of Russia,

with notices of the export of timber from Memel, Dantzig, and Riga

John Croumbie Brown

Forests and Forestry in Poland, Lithuania, the Ukraine, and the Baltic Provinces of Russia, *with notices of the export of timber from Memel, Dantzig, and Riga*

ISBN/EAN: 9783337298890

Printed in Europe, USA, Canada, Australia, Japan

Cover: Foto ©berggeist007 / pixelio.de

More available books at **www.hansebooks.com**

FORESTS AND FORESTRY

IN

POLAND, LITHUANIA,

THE UKRAINE,

AND

THE BALTIC PROVINCES OF RUSSIA,

WITH NOTICES OF THE EXPORT OF TIMBER FROM

MEMEL, DANTZIG, AND RIGA.

COMPILED BY

JOHN CROUMBIE BROWN, LL.D., &c.

EDINBURGH:
OLIVER AND BOYD, TWEEDDALE COURT
LONDON: SIMPKIN, MARSHALL, & CO.,
AND WILLIAM RIDER & SON.
MONTREAL: DAWSON BROTHERS.

1885,

ADVERTISEMENT.

In the Spring of 1877 I published a *Brochure* entitled *The Schools of Forestry in Europe : a Plea for the Creation of a School of Forestry in connection with the Arboretum in Edinburgh*, in which, with details of the arrangements made for instruction in Forest Science in Schools of Forestry in Prussia, Saxony, Hanover, Hesse, Darmstadt, Wurtemburg, Bavaria, Austria, Poland, Russia, Finland, Sweden, France, Italy, and Spain, and details of arrangements existing in Edinburgh for instruction in most of the subjects included amongst preliminary studies, I submitted for consideration the opinion 'that with the acquisition of this Arboretum, and with the existing arrangements for study in the University of Edinburgh, and in the Watt Institution and School of Arts, there are required only facilities for the study of what is known on the Continent as Forest Science to enable these Institutions conjointly, or any one of them, with the help of the others, to take a place amongst the most completely equipped Schools of Forestry in Europe, and to undertake the training of foresters for the discharge of such duties as are now required of them in India, in our Colonies, and at home.'

At a meeting held on the 28th March 1883, presided over by the Marquis of Lothian, it was resolved 'that it is expedient in the interests of Forestry, and to promote a movement for the establishment of a National School of Forestry in Scotland, as well as with the view of furthering and stimulating a greater improvement in the scientific management of woods in Scotland and the sister countries, which had manifested itself during recent years, that there should be held in Edinburgh, during 1884, and at such season of the year as may be arranged, an International Exhibition of Forest Products and other objects of interest connected with Forestry.'

At a large and influential meeting held within the Forestry Exhibition on the 8th of October 1884, it was resolved to establish in Edinburgh a National School of Forestry and a Museum connected therewith, and a Committee was appointed to carry out the resolution.

In a note appended to a circular issued by this Committee it was stated :—'The Committee feel a difficulty in suggesting a definite scheme for the proposed School of Forestry, until they have some knowledge of the amount of funds which may be placed at their disposal; but, in the meantime, it may be sufficient to state that they contemplate the establishment of a Professorship of Forest Science, for the instruction of students in all that pertains both to Practical and Scientific

Forestry—including the physiology and pathology of trees, the climatic and other effects produced by forests, the different methods of forest management, the economic uses to which forest products have been or may be applied, forest engineering, and forest administration generally. Instruction to be communicated by lectures, examinations, excursions, &c.

' A large collection of forest implements, produce, and specimens, was, at the close of the Forestry Exhibition, placed at the disposal of the Committee for use in connection with the proposed Forest School ; and it is the intention of the Committee, should their funds permit, that that this collection should, with such additions as may from time to time be available, be placed in a permanent Museum in connection with the School of Forestry.'

In furtherance of the same object the following volume has been compiled for publication.

JOHN C. BROWN.

HADDINGTON, 15th *May*, 1885.

CONTENTS.

CONTENTS.

PAGE

PART V.—Baltic Provinces of Russia.

AUTHORITIES CITED.

Anderson, pp. 99, 200, 215; Boerling, p. 201; Von Berg, p. 18; Bektiff and Khvostoff, p. 207; Bitney. pp. 37, 53; Bonifacy, p. 64; Campbell-Walker, pp. 29, 33; Dixon, p. 227, 230; Duglossius, p. 68; *Edinburgh Encyclopaedia*, p. 80; Des Fontaines, pp. 66, 80; *Gazetteer*, pp. 93, 243; Hove, p. 53; Jozef, p. 65; Kauffmann, p. 253; Krausse, p. 17; Krysztoff, p. 64; *Le Marchand du Bois*, p. 255; Leroy-Beaulieu, p. 206; *Letters from the Shores of the Baltic*, p. 269; Marny, pp. 35, 227; Mackenzie Wallace, pp. 3, 94, 97; Michie, p. 90; Nauman, p. 14; Novikoff, p. 206; Ostenstacken, p. 60; *Pall Mall Gazette*, pp. 212, 213; Pinkerton, p. 115, 233; Polenjansky, p. 40; Potujanski, p. 64; Polytaief, pp. 129-190; Quinn, pp. 235, 241, 246, 248, 257; *Russian Government Reports*, pp. 36, 60, 115, 116, 119, 120, 123, 191, 229, 261, 262, 264, 266; *Timber Trades Journal*, p. 194; Vidomosti, p. 199; Waga, p. 64.

FORESTRY IN POLAND, LITHUANIA,

AND THE

BALTIC PROVINCES OF RUSSIA.

———o———

INTRODUCTION.

IN the Empire of Russia may be studied many of the phases of forest economy.

In a volume entitled *Introduction to the Study of Modern Forest Economy*, and in a companion volume entitled *Finland : its Forests and Forest Management*, is embodied information in regard to what is there called *Svedjande*, known in French forest science as *Sartage*, and in some parts of India as *Koomaree*.

In another companion volume, entitled *Forest Lands and Forestry of Northern Russia*, is embodied information in regard to exploitation according to what is known in France as *Jardinage*.

In a third volume, entitled *Forestry in the Mining Districts of the Ural Mountains in Eastern Russia*, information is supplied in regard to what is known as *Furetage*, and in regard to what is known as exploitation according to *La Methode à Tire et Aire*, carried on with malversations and abuses, such as in the middle of the seventeenth century called forth a famous and oft-cited warning from Colbert, ' *La France perira en faute des bois !* '

And in a volume entitled *French Forest Ordinance of 1669; with Historical Sketch of Previous Treatment of Forests in France*, is supplied information in regard to stringent measures adopted in France to suppress such malversations and abuses,

B

In Poland we find an endeavour made to introduce the most advanced forest exploitation of the day—that known in Germany, where it originated, as *Die Fachwerke Method*, known in France as *La Methode des Compartiments*, and known in Poland and in some other countries as *the Scientific Method of Exploitation*. In Lithuania we find forest management similar to what prevails throughout Central Russia. And in Courland, Estonia, and Livonia, we meet with some special regulations issued for the management of forests in the Baltic Provinces of Russia.

Through all of these last-mentioned countries—Poland, Lithuania, Courland, Estonia, and Livonia—I have travelled once and again in proceeding to or from St. Petersburg, where I once resided as pastor of the British and American Church, and where I have frequently passed the summer ministering to the same church, while one and another of those who have succeeded me in the pastorate sought a few months' relaxation at home. In all of them I have had correspondents—in some districts, numerous correspondents; and though my correspondence with most related chiefly, if not exclusively, to the publication of religious tracts, and to the sale and distribution of the New Testament Scriptures, there has thus been sustained for more than half a century an interest in them which has procured for me information which otherwise it might have been difficult to obtain.

My first journey through this portion of Western Russia was made in 1836; a second in 1873, coming from Vienna through Poland to St. Petersburg; my last in 1878, proceeding from St. Petersburg through Berlin and Dresden to Paris. In the lengthened interval how changed the mode of travelling! At the time of my first journey there were not even diligences by which the journey from the capital might be made, and railways were unknown and unthought of, while in winter a voyage by sea was impossible.

By enquiry I learned that there was a British courier

likely to leave St. Petersburg shortly for London, going by Berlin, who was quite agreeable to allow me to travel with him for a reasonable consideration; but the time of his departure did not depend upon himself; nor did the route he should follow; and time was of importance to me. I then heard of a merchant going to Riga, desirous of some one to share with him the expense of travelling post; and our arrangements were soon made.

A sledge made of plane deal, with wooden runners, and a canvas curtain in front, was purchased, a *padorozhnaya*, or order for post horses was procured, and off we set. Mr Mackenzie Wallace gives the following account of the posting arrangements of Russia:—

'However enduring and long-winded horses may be, they must be allowed sometimes, during a long journey, to rest and feed. Travelling with one's own horses is therefore necessarily a slow operation, and is already antiquated. People who value their time prefer to make use of the Imperial Post-organisation. On all the principal lines of communication there are regular post-stations, at from ten to twenty miles apart, where a certain number of horses and vehicles are kept for the convenience of travellers. To enjoy the privileges of this arrangement, one has to apply to the proper authorities for a "Podorozhnaya"—a large sheet of paper stamped with the Imperial Eagle, and bearing the name of the recipient, the destination, and the number of horses to be supplied. In return for this document a small sum is paid for imaginary road repairs; the rest of the sum is paid by instalments at the respective stations. Armed with this document, you go to the post-station and demand the requisite number of horses. Three is the number generally used, but if you travel lightly, and are indifferent to appearances, you may modestly content yourself with a pair. The vehicle is a kind of Tarantass, but not such as I have elsewhere described. The essentials in both are the same, but those which the Imperial Government provides resemble an enormous cradle on

wheels, rather than a phaeton. An armful of hay spread
over the bottom of the wooden box is supposed to play
the part of cushions. You are expected to sit under the
arched covering, and extend your legs so that the feet
lie beneath the driver's seat ; but you will do well, unless
the rain happens to be coming down in torrents, to get
this covering unshipped, and travel without it. When
used, it painfully curtails the little freedom of movement
that you enjoy, and when you are shot upwards by some
obstruction on the road, it is apt to arrest your ascent by
giving you a violent blow on the top of the head.

'It is to be hoped that you are in no hurry to start,
otherwise your patience may be sorely tried. The horses,
when at last produced, may seem to you the most
miserable screws that it was ever your misfortune to
behold ; but you had better refrain from expressing your
feelings, for if you use violent, uncomplimentary language,
it may turn out that you have been guilty of gross
calumny. I have seen many a team composed of animals
which a third-class London costermonger would have
spurned, and in which it was barely possible to recognise
the equine form, do their duty in highly creditable style,
and go along at the rate of twelve or fourteen miles an
hour, under no stronger incentive than the voice of the
Yemstchik.   Indeed, the capabilities of these lean,
slouching, ungainly quadrupeds are often astounding
when they are under the guidance of a man who knows
how to drive them. Though such a man commonly
carries a little harmless whip, he rarely uses it except by
waving it horizontally in the air. His incitements are
all oral. He talks to his cattle as he would to animals of
his own species—now encouraging them by tender, cares-
sing epithets, and now launching at them expressions of
indignant scorn. At one moment they are his "little
doves," and at the next they have been transformed into
"cursed hounds."   How far they understand and
appreciate this curious mixture of endearing cajolery and
contemptuous abuse it is difficult to say, but there is no

doubt that it somehow has upon them a strange and powerful influence.

'Any one who undertakes a journey of this kind should possess a well-knit, muscular frame and good tough sinews, capable of supporting an unlimited amount of jolting and shaking; at the same time, he should be well inured to all the hardships and discomfort incidental to what is vaguely termed " roughing it." When he wishes to sleep in a post-station he will find nothing softer than a wooden bench, unless he can induce the keeper to put for him on the floor a bundle of hay, which is perhaps softer, but on the whole more disagreeable than the deal board. Sometimes he will not get even the wooden bench, for in ordinary post-stations there is but one room for travellers, and the two benches—there are rarely more —may be already occupied. When he does obtain a bench, and succeeds in falling asleep, he must not be astonished if he is disturbed once or twice during the night by people who use the apartment as a waiting-room whilst the post-horses are being changed. These passers-by may even order a Samovar, and drink tea, chat, laugh, smoke, and make themselves otherwise disagreeable, utterly regardless of the sleepers. Then there are the other intruders, of which I have already spoken when describing the steamers on the Don. I must apologise to the reader for again introducing this disagreeable subject. Æsthetically it is a mistake, but I have no choice. My object is to describe travelling in Russia as it is, and any description which did not give due prominence to this species of discomfort would be untrue—like a description of Alpine climbing with no mention of glaciers. I shall refrain, however, from all details, and confine myself to a single hint for the benefit of future travellers. As you will have abundant occupation in the work of self-defence, learn to distinguish between belligerents and neutrals, and follow the simple principle of international law, that neutrals should not be molested. They may be very ugly, but ugliness does not justify assassination. If, for

instance, you should happen in awaking to notice a few
black or brown beetles running about your pillow, restrain
your murderous hand! If you kill them you commit an
act of unnecessary bloodshed; for though they may play-
fully scamper around you, they will do you no bodily
harm.

'The best lodgings to be found in some of the small
provincial towns are much worse than the ordinary post-
stations. To describe the filthiness and discomfort of
some rooms in which I have had to spend the night
would require a much more powerful pen than mine; and
even a powerful writer in entering on that subject would
involuntarily make a special invocation for assistance to
the Muse of the Naturalistic school.

'In the winter months travelling is in some respects
pleasanter than in summer, for snow and frost are great
macadamisers. If the snow falls evenly there is for some
time the most delightful road that can be imagined. No
jolts, no shaking, but a smooth, gliding motion, like that
of a boat in calm water, and the horses gallop along as if
totally unconscious of the sledge behind them. Unfortunately, this happy state of things does not last long. The
road soon gets cut up, and deep transverse furrows are
formed. How these furrows come into existence I have
never been able clearly to comprehend, though I have
often heard the phenomenon explained by men who
imagined they understood it. Whatever the cause and
mode of formation may be, certain it is that little hills
and valleys do get formed, and the sledge, as it crosses
over them, bobs up and down like a boat in a chopping
sea, with this important difference, that the boat falls into
a yielding liquid, whereas the sledge falls upon a solid
substance, unyielding and unelastic. The shaking and
jolting which result may readily be imagined.

'There are other discomforts, too, in winter travelling.
So long as the air is perfectly still, the cold may be very
intense without being disagreeable; but if a strong head
wind is blowing, and the thermometer ever so many

degrees below zero, driving in an open sledge is a very disagreeable operation, and noses may get frostbitten without their owners perceiving the fact in time to take preventive measures. Then why not take covered sledges on such occasions? For the simple reason that they are not to be had; and if they could be procured, it would be well to avoid using them, for they are apt to produce something very like sea-sickness. Besides this, when the sledge gets overturned, it is pleasanter to be shot out on to the clean, refreshing snow, than to be buried ignominiously under a pile of miscellaneous baggage.

'The chief requisite for winter travelling in these icy regions is a plentiful supply of warm furs. An Englishman is very apt to be imprudent in this respect, and to trust too much to his natural power of resisting cold. To a certain extent this confidence is justifiable, for an Englishman often feels quite comfortable in an ordinary great coat, when his Russian friends consider it necessary to envelope themselves in furs of the warmest kind; but it may be carried too far, in which case severe punishment is sure to follow, as I once learned by experience. I may relate the incident as a warning to others.

'One day in the winter of 1870-71 I started from Novgorod, with the intention of visiting some friends at a cavalry barracks situated about ten miles from the town. As the sun was shining brightly, and the distance to be traversed was short, I considered that a light fur and a *bashlyk*—a cloth hood which protects the ears— would be quite sufficient to keep out the cold, and foolishly disregarded the warnings of a Russian friend who happened to call as I was about to start. Our route lay along the river due northward, right in the teeth of a strong north wind. A wintry north wind is always and everywhere a disagreeable enemy to face; let the reader try to imagine what it is when the Fahrenheit thermometer is at 30° below zero—or rather let him refrain from such an attempt, for the sensation produced cannot be imagined by those who have not experienced it. Of course I ought

to have turned back—at least, as soon as a sensation of
faintness warned me that the circulation was being
seriously impeded—but I did not wish to confess my
imprudence to the friend who accompanied me.   When
we had driven about three-fourths of the way, we met a
peasant woman, who gesticulated violently, and shouted
something to us as we passed.   I did not hear what she
said, but my friend turned to me and said in an alarming
tone—we had been speaking German—" Mein Gott! Ihre
Nase ist abgefrohren !"   Now the word " *ab*gefrohren," as
the reader will understand, seemed to indicate that my
nose was frozen *off,* so I put up my hand in some alarm to
discover whether I had inadvertently lost the whole or
part of the member referred to.   So far from being lost
or diminished in size, it was very much larger than usual,
and at the same time as hard and insensible as a bit of wood.

' " You may still save it," said my companion, " if you
get out at once and rub it vigorously with snow."

' I got out as directed, but was too faint to do anything
vigorously.   My fur cloak flew open, the cold seemed to
grasp me in the region of the heart, and I fell insensible.

' How long I remained unconscious I know not.   When
I awoke I found myself in a strange room, surrounded by
dragoon officers in uniform, and the first words I heard
were, " He is out of danger now, but he will have a fever."

' These words were spoken, as I afterwards discovered,
by a very competent surgeon ; but the prophecy was not
fulfilled.   The promised fever never came.   The only
bad consequences were that for some days my right hand
remained stiff, and during about a fortnight I had to
conceal my nose from public view.

' If this little incident justifies me in drawing a general
conclusion, I should say that exposure to extreme cold is
an almost painless form of death, but that the process of
being resuscitated is very painful indeed—so painful, that
the patient may be excused for momentarily regretting
that officious people prevented the temporary insensibility from becoming " the sleep that knows no waking."

'Between the alternate reigns of winter and summer there is always a short interregnum, during which travelling in Russia by road is almost impossible. Woe to the ill-fated mortal who has to make a long road journey immediately after the winter snow has melted; or, worse still, at the beginning of winter, when the autumn mud has been petrified by the frost, and not yet levelled by the snow!'

My journey was made in winter, and therefore we had a sledge—a covered sledge, but covered with bass-matting, and all the appointments of the most unpretending character, for it had to be sold at the end of the journey, when it would not fetch a high price. It cost little; but it would sell for less: and a reduction of 50 per cent. on twenty shillings is a good deal less than a like reduction on twenty pounds; we would also be more likely to find a purchaser for our vehicle if it were of little value; and it suited the convenience of both of us to travel cheaply, taking two horses instead of three. On arriving at the first station-house we had to exercise the patience of which Mr Wallace speaks. All the horses were out; and we had to wait till some had returned and rested. It was dark. I strolled about the station-house. Feeling hungry, I looked about for my fellow-traveller to get him to order supper for me, as I was but a novice in travelling there. I could find him nowhere; so I went to lie down in the sledge for a change. Lifting the curtain, I found him there busy making his supper on provisions he had brought with him. I said more to myself than to him, laughing as I said it:—'Holloa, my lad! if you can do that, I can do that too.' And I got out a basket filled with provisions with which some of my friends in St. Petersburg had supplied me. The mention of this recalls the scene of our starting, and one kind face among others now no more. The owner of that face, caring for me as a mother for her son, brought bread, a cake, two tongues, salt—an article most valuable at an *al fresco* meal, but

very likely to be forgotten, and a slab of frozen mince-
collops. I was astounded, but I was assured all would be
needed before I had completed my journey. And so it
was. Morning, noon, and night, when resting for a little
at a station-house, I would ask for a hatchet, and, breaking off a lump of frozen collops, give it to the people to
heat; and my fellow-traveller was not loth to partake
with me of the savoury viands. Here this could not be
done; but getting out a tongue, which, though cooked, had
not been frozen, I made signs to my fellow traveller that
I did not know what to do as I had no knife. Taking the
tongue in one hand, and taking hold of the tip of it with
the other, he screwed this off and held it out to me. I
received it with a laugh; it was my first lesson in rough
travelling; I have travelled many thousand miles since,
in America, in Africa, and in Europe, but never again
have I been at a loss what to do to prepare provisions, and
to use them when prepared. I can kindle a fire in the
desert, and prepare a *carabonaje* with the best of them;
and eat it with a relish, with no garnishing but salt and
pepper. Only once was I placed in any difficulty—and
that not from any squeamishness on my part, but from
the delicacy of feeling experienced by others. Three of
us, with attendants, were crossing the Karoo in South
Africa; there was a stretch of eighty-four miles which
had to be travelled one day. We had purchased a live fowl
at a farm at which we slept the night before. At mid-day,
having out-spanned the horses, the drivers were sent with
them away some six miles to get water. Neither of my
companions could kill a fowl; I volunteered to kill, and
strip, and cook, and eat it too, if they would allow me.
But they shrunk from letting a minister kill the fowl, and
from this delicacy of feeling on their part, we were likely
to have to wait for our dinner till the horses returned,
were in-spanned, and ready to take us forward on our
journey, lest we should be benighted in the *feldt.* We got
out of the difficulty at last by one of the party, not I,
pulling off the head with all his might, and throwing it

away in the air as if it were a thing polluted. Thanks to
my experience at the Russian station-house, I have never
perished of hunger from want of a knife and a silver fork,
though I may equal the most fastidious in the enjoyment
of all the amenities of life.

We stopped an hour or two at Dorpat, that I might
visit some of the professors there with whom I was
acquainted, and just as we were nearing Konigsberg the
English courier overtook us. With him I travelled to
Berlin, rejoicing in the well-made and well-kept roads of
Prussia, which contrasted greatly with those of Russia,
excepting where these were covered with snow, on which
I found it pleasant to glide along in the sledge, travelling
night and day, sleeping when, and only when, so inclined ;
and though sometimes upset, always falling soft on the
uncrushed snow lining the track—our fur shoobs and
wadded caps helping to make the upset more harmless and
rather pleasurable than otherwise from the excitement.

How different is travelling now! In less than three
days one may travel from St. Petersburg to London.
From St. Petersburg to Berlin one may travel with every
comfort without once leaving the carriage, every convenience being provided. From Vienna to St. Petersburg I
once travelled thus, changing carriages only at Warsaw, and
at Wilna, where we joined the train from Berlin. During a
year which my wife spent in Scotland, while I returned and
remained in St. Petersburg, every interchange of letters
cost 13s 6d, and occupied more than a fortnight, nearly
three weeks in transmission. Now the postage of a letter
is 2½d, and it is conveyed in four days, while a post card
is transmitted and delivered for a penny!

PART I.

POLAND,

———o———

CHAPTER I.

IN journeying from St. Petersburg by railway towards Poland, for some 200 miles we pass over ground which is of a dead level, or almost such, being varied, not by rising grounds, but by marshes and bogs, the dry land being to some extent covered with trees of apparently no great age : a stranger would say trees of some twenty or thirty years growth ; and I might say the same, but passing over the ground at distant periods, I have found them always apparently of the same age as they were when I first saw them ; and this may be really the case, these having disappeared, and those now there being trees by which they have been replaced. They are apparently the scraggy representatives of extensive forests of a former day.

Nowhere are seen forests such as may be seen in travelling in the Governments of Olonetz and Archangel in Northern Russia, and of Moscow, Orel, and others in Central Russia.

Passing on, between Pskoff and Dunsburg, the country is found to present more of an undulating aspect; instead of stagnant waters, brooks and rivulets and other forms of running waters are seen. In this district great quantities of flax are grown; and the water is turned to account in preparing the produce for the market. There are two qualities of flax prepared, each in its own way. The one

of these, known as *Motchenets*, is prepared by being steeped
for some time, and dried, to facilitate the removal of the
skin or bark, technically called silica, and other material
from the long woody fibre of which flax consists; the other
of these, called *Slanetz*, is bleached by exposure to rain
and dew, or artificial sprinkling of water, and the sunshine.
The former is white, the latter is dark in colour; and the
latter frequently gives trouble to the merchant by heating
through fermentation about the season at which the plant
flowers.

At Kovno, about 200 miles beyond Dunsburg, we enter
Poland, and advancing through the eastern portion of that
country, the traveller remarks that agriculture appears to
be carried on with more of a scientific character than in
the lands through which he has been passing, both on the
further and on the hither side of Dunsburg; the houses
are more regularly built; the villages have more of a
European aspect, but the houses resembling more those of
Austria than those of Northern Germany, and the similarity may be traced also in the laying out of the fields :
but by this one is reminded more of the arable lands of
Bavaria and other countries of Southern Germany than of
what is seen in Austria.

In North Germany the land is level, and there is no
end of ditches or open drains. Here it is more undulating, and these are less frequent. Agriculture seems also
to be more remunerative than in the lands traversed, the
crops stand thicker in the ground, and surface draining
appears to be uncalled for. The fields are sown with
wheat, whereas to the north of Kovno, there were to be
seen only barley, oats, and flax. All the more valuable
cereals seem to flourish in Poland, and in passing through
this district there is produced an impression that the soil
is more productive than it is further to the north; that
the climate must be more equable; and the superficial
aspect of the land being more undulating, and at the same
time more thickly wooded, that as an agricultural district

it must be at least 50 per cent. superior to the Governments of St. Petersburg and of Pskoff.

In Poland both wheat and wool are raised for exportation. Large crops of potatoes are raised for the production of spirits by distillation, and beetroot for the manufacture of sugar; and wood for building purposes is exported largely. The Scots fir (*Pinus sylvestris*), and the oak (*Quercus robur*), are of very superior quality.

In this district we also find the trees to be different in kind from what they were in the region traversed in coming hither. While in the first stages of the journey they were chiefly and almost exclusively firs, and birches, and willows, here, around Berdicheff, in Poland, we find the woods composed in a great measure of oaks, and elms, and chestnuts; and the forest aspect is completely different.

In the south-east corner of Poland, and in the adjacent districts of Russia known as Little Russia, there are considerable stretches of forest land overgrown by a wild pear tree. The fruit is not edible, either when green or ripe ; but it is gathered and steeped in the Russian beverage called *quass*—the common drink of the peasantry—and it is sold extensively throughout Russia as *steeped pears*, and is greatly in demand by the peasants.

Quass is prepared by pouring hot water upon broken, dry, black rye bread, a little yeast is added, and in some cases a little peppermint, and it is allowed to ferment.

Herds of swine are in the summer time turned out into the woods, where they become little better than wild boars. In November they are driven home, killed and frozen, and sent to the northern districts of Russia to be sold as frozen pork.

By Naumann, in his *Geognosie*, ii. p. 1173, it is stated : ' Olkuez and Schiewier, in Poland, lie in two sand deserts, and a boundless plain of sand stretches around Ozeustackaur, on which there grows neither tree nor shrub. In heavy winds this place resembles a rolling sea, and the sand hills rise and disappear like the waves of the ocean. The heaps of waste from the Olkuez mines are covered with sand to the depth of four fathoms.'

So far as is known to me, no attempts have yet been made to fix and utilise these drifting sands : but that they, as well as those which have been arrested and rendered productive in other lands—Hungary, France, Spain, Portugal, Holland, Denmark, and Northern Germany, may be brought under control I cannot doubt.

But in passing through the country one sees more of forests than of sand drifts.

Of Poland an anonymous writer early in the present century tells :—' Poland is an extremely level country, diversified by few or no eminences, except a ridge of hills branching off from the Carpathian mountains, which anciently formed the southern boundary of the country. The rivers are unadorned with banks, and flow lazily in a flat monotonous course, insomuch that when, as previously stated, heavy falls of rain take place, the country for many miles is completely inundated. The number and extent of marshes and forests, neither of which the Poles have seemed very anxious to remove, uniformly strike strangers as one of the great characteristics of Poland. The soil, which is chiefly either of a clayey or marshy description, is, in many places, so extremely fertile, that with the least cultivation it is calculated to produce the most luxurious crops of corn ; and it is distinguished for the richest pastures in Europe. Agriculture with the Poles, however, is completely in its infancy. For many ages they neglected this useful art, as they neglected every art of peace and domestic comfort; they were a warlike people; and, besides, the produce of the fields was not the property of the peasants, but of their *masters*, and they were themselves doomed, without hope of advancement, to continue in the same rank of life, whatever had been their industry or their skill. But though these disabilities have now been greatly removed, though the Poles are rapidly emerging from that state of laziness and inactivity in which they remained so long sunk, yet, in the department in question they have nearly everything to learn. Of the use of manure they are almost entirely ignorant; their common practice is to crop a field till it be exhausted, and then for a few years to abandon it. Their ploughs are scarcely sufficient to penetrate the *surface* of the ground; and their fields when reaped, exhibit from this circumstance, as rich a verdure as if they had remained for years unbroken. This ignorance, however, is diminishing every day. Some portions of Poland have been denominated the garden of Europe; and a period may not be far distant when the term may, with much propriety, be applied to the whole territory. Societies for the encouragement of agriculture have been established in Poland; and the vast tracts of forests and marshes with which it abounds certainly open up on extensive field for the display of skill and enterprise.'

In 1881 there was published in Warsaw, in the Russian language, a volume entitled *A Critical Examination of the Forest System in the Kingdom of Poland*, by A. Krause, teacher in the College of Agriculture and Forestry in Neu-Alexandria. In this he gives the history of forests and woodcraft in Poland from the earliest times to the present day, with many curious extracts from original documents and authorities. The first period comprises heathen times, when forests were surrounded by mystery and superstitious observances.

The second period extends from the introduction of Christianity in A.D. 860, to the extinction of Polish independence in 1796. In the earlier centuries Poland seems to have been almost covered with immense masses of pine, oak, beech, spruce, lime, larch, and yew. These forests swarmed with wild bison, beavers, wild horses, wild boar, red deer, elk and roe deer, bears, wolves, and lynxes. In the fourteenth century, however, man seems to have reclaimed much of the wilderness. Forest statutes were published by Casimir the Great in 1347. A few titles from this ordinance are :

(1.) *De his qui in silvis alienis damna faciunt*, concerning persons doing damage in the woods of others.

(2.) *De incendentibus silvas vel gaues alienas*, concerning persons cutting down the woods or *field trees* of others.

(3.) *Si alienos porcos in tua silva reperias*, if thou findest another man's swine in thy wood.

(4.) *De silvis glandariis et faguriis*, concerning woods for the production of oak-mast and beech-mast.

(5.) *De incendiarus*, concerning fire raising.

C

' Between 1796 and 1807 the separation of forests into
blocks was commenced. And in 1808 the French civil
code, with its enactments regarding the working of forests
was introduced.'

In 1858 Baron von Berg, Oberforstrath in Saxony, was
applied to professionally to examine and report on the
state of the forests and of forest management in Finland.
Thereafter he was applied to to do the same in
Poland; and he embodied his views in *Denkschrift über
das Forst-Wesen in Polen*, which was published in Leipsic
in 1864 or 1865. Thereafter there was introduced into
the management of the forests the German methods of
treatment, but, as a forest official remarked to me with a
shrug, the principles are the same; but the application of
them, controlled by circumstances, soil, situation, climate,
forest products, &c., is one thing in Germany, it is a very
different thing in Poland, and necessarily so. According
to Herr Krause the estimates for all forest operations
must be prepared some years beforehand. Every forest
must be surveyed, and charts of it mapped out to a scale
of $1 = 20,000$, in which one English mile would be
represented by about an inch. Smaller plans must also
be prepared to a scale of three miles to an inch. A block
coincides, as a rule, with the district perambulated by one
under-forester. Each block is divided into four divisions,
and each division is subdivided into fifteen compartments.
In each compartment all the trees are intended to be as
nearly as possible of one age or varying by not more than
thirty years. An interval of thirty years is allowed after
maturity for seeding and ensuring natural reproduction.
Each division, if perfect according to plan, should contain
fifteen different classes, or ages, of timber. This implies
for each division, as also for the whole block or forest, that
each tree should be felled between the ages of 120 and
150 years. The rotation of felling, instead of being
directed, as is most usual, from east to west, proceeds in
regular order in each division, from south-east towards the

north-west. The avenues and paths bounding the compartments, run in parallel straight lines from north-east to south-west, with the other avenues intersecting them not quite at right angles. The compartments have been almost always made equal in area, without taking account of the difference in the qualities of soil. Each compartment (according as the forest surveyor shall direct in his estimates) must either be parcelled out into thirty yearly portions, or, its cubic contents having been measured, and its rate of increase ascertained, equal solid quantities of timber is felled in each year of the thirty. In general the latter method is prescribed as being more favourable to the ripening of seed, and the ripening of seedlings with little artificial assistance. Compartments occupied by coppice or by coppice with reserved trees (compound coppice) are parcelled out into equal yearly portions.

In criticising this Polish system, and awarding it in its place among scientific systems, Herr Krause, after a multitude of definitions and extracts, comes to the conclusion that it is a high development of the method of equal yearly clearings. Its descent he traces from the system which prevailed in Brunswick at the commencement of this century. (It is said that the system of equal yearly portions was first devised by Frederick the Great of Prussia.) And he states that by a series of special alterations in the direction of completeness, and some improvements which he suggests, the Polish system of estimates may readily be brought up to the most modern form of methodical exploitation.

The method of exploitation referred to is known in France, as has been stated, as *La Methode des Compartiments*, in Germany as *Die Fachwerke Method*, here as the *Scientific Method*.

The system of *Jardinage*, felling a tree here and there as it might be wanted, was in France and other countries on the Continent of Europe extensively superseded more than two hundred years ago by exploitation *à tire et aire*, in which all the trees on a given area were felled with the exception of some left to bear seed, and to afford shade and shelter to seedlings. The areas so exploited frequently succeeded each other side by side in regular progression. Under the scientific method, the *Fachwerke Method*, or *Methode des Compartiments*, plots or patches of forest like to each other, but situated apart, are treated as if they formed a continuous wood, and are treated collectively as, were the different areas in exploitation *à tire et aire*.

Where *Jardinage* is followed we have trees of varied ages growing confusedly mixed together, and injuring one another, the older and taller overgrowing the younger and impeding their growth, and the older becoming straggling and knotty, and of a height inferior to what they attained when they were more serried, while they are also less able to withstand the storm; and many, young and old alike, often become diseased, rarely appear well conditioned, and not unfrequently perish prematurely. Though some trees withstand and surmount all these evils, the product of a forest so treated is in a given time inferior both in quantity and in quality to what it might have been; and there is a tendency in the method of management to convert the forest into a scrub, and ultimately to destroy it altogether.

About the middle of the sixteenth century it began to be realised that the system followed had in it inherent defects, and these such as could scarcely be eradicated by any attempted improvement to which it could be subjected. But it was thought that the conservation of the forests might be assured by continuous reproduction if the fellings were confined to one portion of the forest, throughout one year, or a series of years, more or less prolonged, during which the trees growing elsewhere should be allowed undisturbed growth; while this should be allowed to recover itself during the much more lengthened period which would be required to go over not one but all of the other divisions of the forest or woodlands with which it was connected, and if necessary the extension of this to other forests or other woodlands.

Such was the *Methode à tire et aire*, associated by many with the celebrated Ordinance of 1669 ; but sustained production was not thereby secured. Forests continued to disappear, and though this was attributable to abuses for which the system was not responsible, it was remarked that there was a great inequality in the products of successive fellings during successive periods ; that there was a considerable loss of possible produce not only from difference in the productiveness of different patches in the same time, but more so from the circumstance that as some portions were necessarily cut down when the trees were too young, in other portions many trees, and even entire plots, were left standing so long waiting for their turn that they decayed before they came within the regular series of fellings.

The succeeding crops, moreover, were not always found equal in quantity to the preceding, while they were also much inferior in quality in consequence of the irregularity in regard to denseness in which they grew up—in some places sparse, in others so dense that they could neither attain to good proportions nor acquire a firm texture. These evils it was sought to remedy by giving to every patch or plot in the forest the treatment it specially required, combining those which were in like conditions and conveniently situated, treating them as if they constituted a continuous wood —modifying the arrangement of *à tire et aire* to meet the requirements of this development of that system, and adding the products of thinnings and partial fellings in patches or plots, in which these were practised, to the products of the successive definitive fellings in others, or rather adding these to those to complete the supply required without preventing natural reproduction and sustained production.

In the accomplishment of this there are sundry operations necessary in any case, but more especially when it is desired to secure all the advantages of the system. Circumstances may determine the order in which these may be attended to, and it appears to me to be a matter of indifference in what order I now detail them.

Let it be supposed that it is virgin forest which is to be exploited—it is ascertained, by survey and inspection, of what kinds of trees it is composed; what age, or different ages, the different kinds of trees are; in what condition they are; what measure of vigour of growth they manifest, and what are the promises, in regard to this measure of vigour being maintained or increased in coming years, supplied by the nature of the soil and the situation of the trees. Thus a general idea of the *peuplement* of the forest is obtained. By another series of observations which may have previously been made in connection with, or in preparation for, the exploitation of other woods, or, which if not previously made, must be made now, it must be ascertained in regard to the kind -or different kinds of trees of which the *peuplement* of the forest consists—what increase of cubic contents these make in the course of a year, or of any definite number of years—say five or ten—at all stages of growth, from that of a sapling to that of an old tree beginning to decay; and in connection with this, by deduction from data thus obtained or by separate observation, it must be ascertained to what age and magnitude the kind of tree growing there may attain without decay; at what age the annual increase becomes so diminished that a greater production of wood in a given period of considerable duration may be obtained by felling and raising a new crop, than by allowing the trees to go on growing; at what age of growth the wood is of best quality for the purpose for which it is required; and at what age a maximum of wood of such superior quality would be secured by then felling the tree.

In illustration of what may thus be learned I may adduce a hypothetical case. Suppose, what is likely to be the case, that a tree in its growth makes an increasing increment of cubic contents year by year, and decade by decade, up to the age of sixty, making more wood between 10 and 20 than between 1 and 10, more between 20 and 30 than between 10 and 20, and so on continuously, more between 50 and 60 than between 40 and 50: it may be found that, though continuing to make increase, it does not make a greater increment between 60 and 70 than it did between 30 and 40, and that, though still making increase, this is now in continuously diminishing quantity, until decay having begun after a lapse of years, it loses more by decay than it makes by growth. But it may be found that though the rate of increase of quantity was reversed after 60 years of growth, the quality continued to improve for thirty years and more thereafter; and also that it was only when the tree had attained the age of 150 years that it had attained the size necessary to yield timber of the size required for some special purpose for which it was needed. All of the information thus obtained may be turned to account in determining the treatment to be given to the forest in exploitation, though it may, or it may not, all be required; but the same may be said of a great deal of the work of subordinates in any work.

In the hypothetical case supposed, it might be necessary to secure some timber of the greatest bulk which could be obtained, and in order to this, that the trees yielding this should grow to the age of 150 years. But it might be desired rather to get as much wood from the forest as possible; and much more would be obtained in the course of 300 years by felling at the age of 75, and getting four crops of trees of that age than by felling all at 150, and getting two crops of trees at that age. By felling at 60, and leaving *baliveaux* for seed, several fellings of trees aged 90—the age at which the trees yielded timber of the best quality—and fellings at the ages of 120 years, and 180 years, and 240 years, and 300 years, of timber of greater magnitude might be obtained, along with the coppice wood of six fellings in the course of 300 years, from *baliveaux* left growing at the first, the second, and the third successive fellings; which might be reserved exactly in such numbers as would yield the timber required; and the rapid growth of none of the others at an earlier age would be sacrificed. Or, again, it might be desired to get as large pecuniary returns from the forest as possible; and this might be secured by an improved quality, though in diminished quantity, and that the maximum quantity of wood of the requisite quality could be secured in the period named—300 years, by felling the trees at the age of 100, and getting three crops of trees of that age off the ground, than by letting all grow to the age of 150; while larger timber required might be obtained, by reserved *baliveaux;* and arrangements might be made for the felling of such of these as might be required, not only without detriment, but if properly managed, with benefit to the crops then growing. These are but a few of the simplest complications for which a director of operations prepares.

The director or projector of operations, with the data referred to before him, together with special requirements such as those just alluded to, if any such there be, determines what period of time is to be assigned to the crops to be produced successively by the forests—in technical phrase, determines the *regimé,* or if this has been pre-determined, determines the time to be allotted to a *revolution,* or cycle, from sowing through growth and successive thinnings to felling and resowing again. The number of years going to a complete *revolution* is very different under the coppice wood *regimé* and under the *regimé* of timber forest; and under each of these it may vary within a comparatively limited range, according to the kind of tree, the situation, the demand, and the vigour of growth. This, then, has to be determined next conditionally, at least, if not absolutely.

With this addition to the data an estimate is then made of what is called *la possibilité,* or of the cubic measurement of the growth made year by year, or definite period of years throughout the period of the *revolution;* which supplies a measurement of the wood, which, without exceeding the yield, might be removed in the course of each of these years, or periods of years.

By withdrawing only the quantity thus determined, the forest may be exploited without imperilling its sustained production, provided the exploitation be equably diffused both as to time and space.

With a view to this there is prepared—and it may have been in course of preparation in the forest while some of these data were being prepared in the office—a survey, trigonometrical or other, of the forest and of all the plots or patches of which it is composed, varying in any way from what is adjoining, with a diagram of the whole representing each of these, accompanied by a report specifying the particulars of each. And with these before him, together with the other data, the forest official entrusted with the work proceeds to what is called *l'assiette des coupes*, the allotting of the different portions to be felled in successive periods, which is done with provision for subsequent rectification, if, through anticipated contingencies or unforseen incidents, this should prove desirable.

Let us suppose the forest under consideration is to be subjected to the *regimé* of timber forest, with a *revolution* fixed at 120 years, it may be divided into four approximately equivalent portions to be exploited successively in four successive periods of thirty years, but in any or all of which there may be carried on extraordinary operations which occasion may call for, the produce of which are reckoned on as a component part of the annual yield of the forest.

The portion allotted for exploitation in the first period of thirty years is then on a like principle subdivided into approximately equivalent portions for exploitation during successive periods of say five years. And there is prepared a provisional scheme of annual operations throughout the first sub-period of five years, which, when sanctioned, are followed for a time; but from time to time the scheme is reviewed, proceeds are compared with estimates, and if necessary the scheme of operations is modified.

These operations include not only the fellings of certain portions, but the thinnings of others repeated from time to time, and the removal of some of the *baliveaux* left for seed, and shelter, and shade, a sufficient number of trees being left in each felling to resow the ground and give shelter to the seedlings. By these reserved trees there is secured the natural reproduction of the forests in connection with the sustained production of the same, and by them shelter required by the seedlings is supplied ; but as these advance in growth their being overtopped would prove a hindrance to their growth. All that is desirable is such shade as might be supplied by the shadow of the tree passing over them as the sun advances in his course ; and all the *baliveaux* excepting such as might sufficient to yield this are removed. Subsequently these also may be removed with benefit to the new *repeuplement*, and they also are removed. To the new *repeuplement* it is also advantageous that from time to time they should be thinned ; and this is done either at the time that *baliveaux* are being felled, or at other times. By these successive operations there is combined with the natural reproduction a progressive amelioration of the forest ; and the products of these operations are taken into account along with the products of the regular fellings in reckoning the periodic produce.

Besides these extraordinary products of thinnings, &c., added to the ordinary products of the regular fellings, it occasionally happens that a storm or a fire occasions such devastation that some, or many, or all of the trees must be cleared away, and an estimated average of the additional wood thus supplied for the market enters into the estimate of the periodic produce.

The skill of the forest agent entrusted with the work is seen in his so apportioning and allotting the fellings in the division assigned to the period, and the thinnings over the whole area of the forest, so as best to secure the three objects aimed at—sustained production, progressive amelioration, and natural reproduction. While this operation supplies ample scope for the exercise of skill, instruction is given in the School of Forestry of the country in regard to the means of doing what is required. The instructions I cite not here, but content myself with reporting what is done in such a way that a definite idea may be formed of what the improved system of exploitation is, and of wherein it differs from *Jardinage*, felling a tree here and there as required, and from *Exploitation à tire et aire*, the progressive felling of adjacent areas.

I have spoken of a virgin forest to be subjected to the forest *regimé*. But there are other cases, some or all of which much more frequently exercise the skill of the forest administrator. It may be required of him to subject such a forest to the coppice wood *regimé*, or a coppice wood to the timber forest *regimé*, or a forest which has been subject to the timber forest *regimé* to the coppice wood *regimé*, or a mixed coppice and timber forest to either a simple timber forest or a simple coppice wood *regimé*, or to convert a timber forest or coppice wood into such a mixed coppice and timber forest, or a forest which has been treated by *Jardinage* into simple coppice, into mixed forest and coppice, or into simple timber forest, or a timber forest or coppice, or one which has been subjected to exploitation according to *La Methode à tire et aire*, into one or other of these conditions—all with a view to subsequent exploitation according to the *Fachwerke Method*, or *Methode des Compartiments*.

All such operations afford scope—more ample scope than does the hypothetical case previously supposed, for the exercise of the forester's skill, but instructions are given in many of the Schools of Forestry—for example, in the school at Nancy, in France—in regard to what should be done in each and all of the cases mentioned ; and these instructions are simply appropriate applications to each case of the general principles an exposition of which has been given.

In the instructions given, as in instructions given in a school of surgery, of medicine, or of law, all that is done, and all that can be done, is to demonstrate and establish general principles, with illustrations of their applicability, and of their application to certain definite cases, instruc-

ting, educating, and training the student, and sending him
forth to exercise his common sense, his judgment, and his
skill, as occasion may demand, and circumstances may
warrant or permit.

Under the antiquated modes of exploitation, *Jardinage*,
and exploitation according to *La Methode à tire et aire*, the
former still practised in several British colonies, the latter
in its most imperfect form in some parts of Russia, and in
some parts of Scotland, explicit rules could be laid
down for the guidance of woodmen and foresters.
Under the more advanced Continental forest economy of
the day general principles of procedure are evolved;
and the teachers of this forest economy, as do the teachers
of law, surgery, and the practice of medicine, say we must
leave the application of our instruction to the skill of the
practitioner. The legal practitioner whose advice is asked
may say, the law is so-and-so; but I must see the documents before I can give an opinion, for, the case being
altered, that alters this case. The medical or surgical
practitioner whose advice is asked may say, the general
principles applicable to the case are so-and-so; but so
much depends on circumstances that some medical man
must see the case before a prescription can be given. And
so is it with the expounders and practitioners of the
advanced forest economy of the day. They say, these are
the principles upon which we proceed, but each particular
case requires its own peculiar application or mode of application of them. And this, which may seem to be to
its disparagement, is considered by many, and considered
justly, to speak its excellence.

In regard to this method of exploitation, which I consider that required in the primitive forests existing in
some of our colonial possessions to ensure their conservation, profitable exploitation, and natural reproduction, I
have given additional illustrations in a volume entitled
French Forest Ordinance of 1669; with Historical Sketch of

*Previous Treatment of Forests in France** (pp. 45-47); and in *Introduction to the Study of Modern Forest Economy†* (pp. 165-186).

In the latter I have embodied the following extracts from a paper read by Captain (now Lieutenant-Colonel) Campbell-Walker before the Otago Institute in Dunedin, New Zealand, on December 21, 1876, entitled *State Forestry: its Aim and Object.* He says in regard to the way in which operations are initiated in Germany and France :—

'When a forest is about to be taken in hand and worked systematically, a surveyor and valuator from the forest staff are despatched to the spot—the former working under the directions of the latter, who places himself in communication with the local forest officer (if there be one), the local officials and the inhabitants interested, and obtains from them all the information in his power. The surveyor first surveys the whole district or tract, then the several blocks or subdivisions as pointed out by' the valuator, who defines them according to the description and age of the timber then standing, the situation, nature of soil, climate, and any other conditions affecting the rate of growth and nature of the crops which it may be advisable to grow in future years. Whilst the surveyor is engaged in demarcating and surveying these blocks, the valuator is employed in making valuations of the standing crop, calculating the annual rate of growth, inquiring into and forming a register of rights and servitudes with a view to their communication, considering the

---

* *French Forest Ordinance of 1669 ; with Historical Sketch of Previous Treatment of Forests in France.*—The early history of forests in France is given, with details of devastations of these going on in the first half of the seventeenth century ; with a translation of the Ordinance of 1669, which is the basis of modern forest economy ; and notices of forest exploitation in *Jardinage*, in *La Méthode à Tire et Aire*, and in *La Méthode des Compartiments*.

† *Introduction to the Study of Modern Forest Economy.*—In this there are brought under consideration the extensive destruction of forests which has taken place in Europe and elsewhere, with notices of disastrous consequences which have followed—diminished supply of timber and firewood, droughts, floods, landslips, and sand-drifts—and notices of the appliances of modern forest science successfully to counteract these evils by conservation, planting, and improved exploitation, under scientific administration and management.

best plan of working the forest for the future, the roads which it will be necessary to construct for the transport of timber—in fact, all the conditions of the forest which will enable him to prepare a detailed plan for future management, and the subordinate plans and instructions for a term of years, to be handed over to the executive officer as his "standing orders." A complete code of rules for the guidance of the valuators has been drawn up and printed, in which every possible contingency or difficulty is taken into consideration and provided for. Having completed their investigations on the spot, the valuator and surveyor return to head-quarters and proceed to prepare the working plans, maps, &c., from their notes and measurements. These are submitted to the Board or Committee of controlling officers, who examine the plan or scheme in all its details, and if the calculations on which it is based be found accurate, and there are no valid objections on the part of communities or individuals, pass it, on which it is made out in triplicate, one being sent to the executive officer for his guidance, another retained by the controlling officer of the division, and the original at the head quarter office for reference. The executive officer has thus in his hands full instructions for the management of his range down to the minutest detail, a margin being of course allowed for his discretion, and accurate maps on a large scale showing each subdivision of the forest placed under his charge.'

With regard to measures adopted to secure natural reproduction of exploited forests, he says:—
' Natural reproduction is effected by a gradual removal of the existing older stock. If a forest tract be suddenly cleared, there will ordinarily spring up a mass of coarse herbage and undergrowth, through which seedlings of the forest growth will rarely be able to struggle. In the case of mountain forests being suddenly laid low, we have also to fear not only the sudden appearance of an undergrowth prejudicial to tree reproduction, but the total loss of the

soil from exposure to the full violence of the rain when it is no longer bound together by the tree roots.   This soil is then washed away into the valleys below, leaving a bare or rocky hillside bearing nothing but the scantiest herbage.   We must therefore note how Nature acts in the reproduction of forest trees, and follow in her footsteps. As Pope writes—

> ' First follow Nature, and your judgment frame
> By her just standard, which remains the same,
> Unerring.' . . . .

Acting on this principle, foresters have arrived at a systematic method of treatment, under which large tracts of forest in Germany and France are now managed.   The forests of a division, working circle, or district, are divided according to the description of the timber and the prevailing age of the trees, and it is the aim of the forester gradually to equalise the annual yield, and ensure its permanency.   With this object, he divides the total number of years which are found necessary to enable a tree to reach maturity into a certain number of periods, and divides his forest into blocks corresponding with each period or state of growth.   Thus, the beech having a rotation of 120 years, beech forests would be divided into six periods of 20 years each—that is to say, when the forest has been brought into proper order, there should be as nearly as possible equal areas under crop in each of the six periods, viz., from 1 year to 20, from 20 to 40, and so on.   It is not necessary that the total extent in each period should be together, but it is advisable to group them as much as possible, and work each tract regularly in succession, having regard to the direction of the prevailing winds.   When a block arrives in the last or oldest stage, felling is commenced by what is called a preparatory or seed clearing, which is very slight, and scarcely to be distinguished from the ordinary thinning carried on in the former periods.   This is followed by a clearing for light in the first year after seed has fallen (the beech seeds only every fourth or fifth year) with the objects of—1st, pre-

paring the ground to receive the seed ; 2nd, allowing the
seed to germinate as it falls ; 3rd, affording sufficient light
to the young seedlings.   The finest trees are, as a rule,
left standing, with the two-fold object of depositing the
seed and sheltering the young trees as they come up.   If
there be a good seed year and sufficient rain, the ground
should be thickly covered with seedlings within two or
three years after the first clearing, Nature being assisted
when necessary by hand sowing, transplanting from
patches where the seedlings have come up very quickly,
to the thinner spots, and other measures of forest craft.
When the ground is pretty well covered the old trees are
felled and carefully removed, so as to do as little damage
as possible to the new crop, and the block recommences
life, so to speak, nothing further being done until the first
thinning.   The above is briefly the whole process of
natural reproduction, which is the simplest and most
economical of all systems, and especially applicable to
forests of deciduous trees.   The period between the first
or preparatory clearing and the final clearing varies from
ten to thirty years, the more gradual and protracted
method being now most in favour, particularly in the
Black Forest, where the old trees are removed so gradually
that there can scarcely be said to be any clearing at all, the
new crop being well advanced before the last of the parent
trees is removed.   This approximates to "felling by selec-
tion," [*Jardinage*], which is the primitive system of working
forests in all countries, under which, in its rude form, the
forester proceeds without method, selecting such timber as
suits him, irrespective of its relation to the forest increment.   Reduced to system, it has certain advantages,
especially in mountain forests, in which, if the steep slopes
be laid bare area by area, avalanches, landslips, and disastrous torrents might result, but the annual output under
this system is never more than two-thirds of that obtained
by the rotation system, and there are other objections
which it is unnecessary to detail in this paper, which have
caused it to be rightly condemned, and now-a-days only

retained in the treatment of European forests under peculiar or special circumstances.'

In a volume entitled *Reports on Forest Management in
Germany*, Colonel Campbell-Walker has said :—' The main
object aimed at in any system of scientific forestry is, in
the first instance, the conversion of any tract or tracts of
natural forests, which generally contain trees of all ages
and descriptions, young and old, good and bad, growing
too thickly in one place and too thinly in another, into
what is termed in German, a *geschlossener Bestand* (close or
compact forest), consisting of trees of the better descriptions, and of the same age or period, divided into blocks,
and capable of being worked, *i.e.*, thinned out, felled, and
reproduced or replanted, in rotation, a block or part of a
block being taken in hand each year. In settling and
carrying out such a system, important considerations and
complications present themselves, such as the relation of
the particular block, district, or division, to the whole
forest system of the province; the requirements of the
people not only as regards timber and firewood, but straw,
litter, and leaves for manure and pasturage ; the geologi-
cal and chemical formation and properties of the soil ; and
the situation as regards the prevailing winds, on which
the felling must always depend, in order to decrease the
chances of damage to a minimum ; measures for precautions against fires, the ravages of destructive insects, tres-
pass, damage, or theft by men and cattle. All these must
be taken into consideration and borne in mind at each
successive stage. Nor must it be supposed that when
once an indigenous forest has been mapped, valued, and
working plans prepared, the necessity for attending to all
such considerations is at an end. On the contrary, it is
found necessary to have a revision of the working plan
every ten or twenty years. It may be found advisable to
change the crop as in agriculture, to convert a hard wood
into a coniferous forest, or *vice versa*, to replace oak by
beech, or to plant up (*unter bau*) the former with spruce or

D

beech to cover the ground and keep down the growth of grass. All these, and a hundred other details, are constantly presenting themselves for consideration and settlement, and the local forest officer should be ever on the alert to detect the necessity of any change and bring it into notice, and no less than the controlling branch should be prepared to suggest what is best to be done, and be conversant with what has been done, and with what results, under similar circumstances, in other districts and provinces.'

Such is the method of exploitation which was introduced into Poland in 1858 by Baron von Berg; and though there may have been such a degeneration by reversion to older and inferior methods of exploitation, as is intimated by Herr Krause in his *Critical Examination of the Forest System of the Kingdom of Poland*, the method in its characteristic details is still followed.

CHAPTER III.

M. MARNY writes of the forests in Poland :—' In Poland
we meet with only a few forests capable of giving any
idea of what was the ancient forest condition of the
country. A sample of this may be seen in the forest of
Wodwosco, which lies upon the domain of that name,
between those of Uraniezko and of Sublowiez. Whilst
one part cleared early in this century offers only a con-
tinuation of bushes and thickets, in the midst of which
spring up here and there a few alders, maples, or hollies;
in that which the hand of man has respected to this day,
the forest offers admirably tall trees of oak and beech,
mingled with majestic firs. Where the bushes disappear
a carpet of moss and heath re-cover the soil. Beyond this
the land loses this uniformity, and becomes more broken;
a torrent dashes with *fracas* over the *débris* of rocks.
The trees are crowded together, and their branches are
drawn nearer and nearer, forming a dome which the rays
of the sun seek in vain to penetrate.'

The annual tabulated reports of the area of forests in
Poland, and in the different divisions of Poland, vary
considerably. This may be attributable in part to one or
other, or one or more, or all of the following causes :—
Extensive clearings, extensive plantings, and rectifications
of estimates by new surveys and accurate measurements.

From a series of these in my possession I give the
following, which gives in a tabulated form for 1870 information which may be generally interesting :—

	Area of Forests.	Area of Crown Forests. (In desatins.)	Extent of Forest to square verst.	Extent of Crown Forests to sq. verst.	Extent of Forests to Inhabitant.	Extent of Crown Forests to Inhabitant.	Annual Revenue from Crown Forests per dec. kopecs.
Warsaw,	301,000	36,037	23·7	2·8	·3	·04	102·3
Kalish,	210,000	19,333	21·7	1·9	·3	·03	47·1
Kielce,	267,000	76,896	31·9	9·2	·7	·2	35·3
Lonija,	276,000	107,512	27·2	10·6	·6	·2	49·3
Lublin,	469,000	14,273	32·4	9·	·7	·02	239·6
Piotrkow,	303,000	77,074	29·1	7·4	·4	·1	56·9
Plotsk,	205,000	27,970	22·3	3·	·4	·06	64·
Radoin,	380,000	81,739	34·7	7·4	·8	·2	37·3
Souvalki,	307,000	109,486	28·7	18·6	·6	·4	16·4
Siedletz,	335,000	18,995	27·4	1·5	·7	·04	15·8

From this it appears that in that year there were 3,053,000 desatins of forest—the desatin measuring 2·69972 Imperial acres. Of these, 569,315 were Crown forests, which gives a mean 27·91 of forests, and 7·14 of Crown forests per square verst—a verst being equal to two-thirds of a British mile ; and 5·5 desatins of forest, 1·29 desatins of Crown forest per inhabitant ; and the mean annual revenue per desatin Crown forests was 66·4 kopecs, equal, at the present rate of exchange, to 1s 2d sterling.

In the Polish Journal of Rural Economy and Forest Science, *Selskoy Ghosyaistvo u Laesovolstvo*, there is given the following information by Mr A. Bitney in regard to the present state of the conservation and exploitation of the Crown forests in Poland:—' The accumulation of a great part of the Crown forests in a few well-known centres leads to the conclusion that they must be very unequally distributed, and, in point of fact, we find the distribution in the ten Governments into which the kingdom is divided to be as follows:—

Governments.	Forests solely belonging to Government.	Proportion to Government Forests. Of Private Forests.	Of Communal Forests.
Warsaw, - -	37,491 dec.	1/6·4	1/36
Kalish, - -	18,870 ,,	1/9·3	1/54
Petrokoff, - -	80,179 ,,	1/2·4	1/14
Radom, - -	116,738 ,,	1/2·10	1/10
Keletz, - -	80,638 ,,	1/2·0	1/11
Lublin, - -	31,024 ,,	1/13·0	1/50
Siedletz, - -	24,927 ,,	1/11·4	1/52
Plotsk, - -	30,445 ,,	1/5·2	1/33
Lonija, - -	97,685 ,,	1/1·5	1/11
Souvalki, - -	192,041 ,,	2/1·0	1/5
Total,	710,038 dec.	$\frac{1}{3}$	$\frac{1}{15}$

' From this it appears that there are five Governments absolutely poor in private forests—Kalish, Siedletz, Lublin, Warsaw, and Plotsk, in which the private forests are in proportion to Government forests as one-ninth, and the communal forests as one forty-third; and there are four Governments—Petrokoff, Keletz, Lonija, and Ladom, in which the private forests are in proportion to the Government forests as one-half, and the communal forests as one-eleventh; and lastly there is the Government of Souvalki, which presents the extreme case of the private forests being to the Crown forests in the proportion of two to one, and the communal forests as one-fifth.

' Such an unequal distribution of Crown forests, with the great dispersion and small extent of a great many of them, places the local forest administration under the disadvantageous necessity of combining several small forest estates in one forest administration, and thus it comes to pass that the different forest administrations in the first mentioned five Governments do not exceed upon an average 7000 desatins, while in the four following they average 13,000 desatins, and in the Souvalki Government 16,000 desatins.

'The whole of the Crown forests of the kingdom are at present classified under three heads :—

'I.—Crown forests in absolute possession, 572,454 desatins in extent, the whole annual produce of which, subject to a small deduction for Government and local requirements may be disposed of by sale.

'II.—Forests under concession to well-known mining works, which are 116,705 desatins in extent, and are assigned to these for the supply of timber, firewood, and charcoal, for which the proprietors or lessees of these works pay into the Treasury a contract price according to the quantity used, while the Forest Administration has a right to dispose according to their judgment of any surplus produce after all the requirements of these works have been met.

'III.—Leased forests, covering an area of 8,383 desatins, in which the wood may be felled by the lessees, but the Government administration is retained.

'From information embodied in a report made by the Central Forest Administration it appears that there is a fourth category—confiscated forests on church lands—covering an area of 12,496 desatins, formerly belonging to clergy of the Romish Church, but which have latterly come under the administration of Government, and which might have been included under the first head, but the arrangement for the administration of them was not then completed, nor were any sales then made.

'Besides these Crown forests, which are under administration, the administrative authorities of the Crown forests are understood to have a surveillance of all those private forests given as security for money loans obtained from the

Government or from banks, and of such as are held by private persons under deed of entail. The forest estates so given in security for loans in all the Governments of the kingdom in 1868 were 417 in number, covering a total area of 162,633, desatins. The entailed forest estates were in number 90, covering an area of 56,854 desatins. In all, 507 forest estates, covering an area of 219,487 desatins.

'The smallest number of forests placed under Government surveillance as forests mortgaged for loans are situated in the Governments of Souvalki, Keletz, and Radom ; the greatest number in the Governments of Petrokoff, Kalish, Olotsk, and Warsaw. The number of entailed estates brought under this surveillance are few in the Governments of Keletz and Radom, with very few indeed in the Governments of Plotsk and Lublin. They are mostly situated in the Governments of Petrokoff, Kalish, Lonija, and Souvalki.

'All the private forests under the surveillance of the Forest Administration are placed under the supervision of their officers, according to their situation and proximity to different forest administrations of Crown forests in the same district; and as there are a great many such forests, and they are widely dispersed over the kingdom, of the 64 administrations amongst which the Crown forests are divided, there are only seven of these which have not private forests under their supervision.

'This supervision extends only to seeing that they be not destroyed. With this in view, and to reduce as much as possible the trouble of the administration in the discharge of the duty, the lenders—be it the Government or the banks—make it a condition of loan that they shall bring if necessary the forests into proper condition, and carry out such plans of management as after examination shall be sanctioned by the Forest Administration. Entailed estates are not handed over to the inheritors until they have been, if necessary, put into a proper condition in accordance with the regulations laid down for Crown

forests, and they bind themselves to carry out the management and exploitation of them in accordance with the pre-
scriptions of these regulations.

' In all that relates to details of conservation—the period
of felling, the mode of sale, and the cutting up of the
wood—the holder is left to do as he thinks proper, provided only a systematic approved exploitation be carried
out, with satisfactory provision for the reproduction of the
forests as they are felled; in regard to these points alone
is surveillance exercised.

' A little consideration will show that the Government
foresters have enough and more than enough on their
hands in the administration of the forests belonging to
the Crown.   And it is not always the case that they have
either the opportunity or the time to carry this out in
accordance with the usually prescribed order of procedure
—and this, exclusive of what has to be done in connection
with partial fellings, and cutting openings through the
extent of the forest, &c., all which makes the control they
have to exercise more complicated, and the result may be
an exhaustion of the forests subjected to their surveillance.
And this is the more likely, seeing that they have no power
in many cases to compel the proprietors to keep a forester
to carry out the prescribed plans of management and exploitation.

' From the consideration of all this it will be seen that
the supervision of private forests is not so easy a matter as
at first sight it may appear to be.   And such supervision
as is exercised cannot prevent exhaustion, or even up-
rooting, or destruction of private forests, if the proprietors
wish this done.

' In regard to the failure of this nominal surveillance, Mr
Polonjansky, in statistics prepared by him some fifteen
years ago, pointed out to the Forest Administration that
up to that time nothing effectual had been done.

' All the Crown forests have been under the supreme
control of the Administration of Finance from the initia-

tion of regular forest administration up to the present time. Under the Minister of Finance there has been a distinct Department for the management of the Crown property, included in which are the Crown forests. Some few changes have taken place in this department, but they have consisted mainly in reducing the number of officials employed in it. At present the forest section of the Administration of Finance consists of thirteen officials in all, under a chief, but even he has very little to do with the actual management of the forests, and he in turn is under the immediate direction of the department of Imperial domains, with the director in which lies the initiation of the management of the forests; and the forest section in the Ministry of Finance presents more the appearance of a chancery than of a department.

'The forests are arranged in eight districts, over each of which is a commissioner; but these districts are not all of equal size and importance. Each commissioner has an adjunct, who may be likened to the taxator, or valuer, in the Russian forest service. They have the superintendence of all buildings, &c., as assistants to the commissioners, and they very rarely have anything entrusted to them entirely.

'The pay of the officials is very moderate. The chief of the section receives 1350 roubles a year. The *Stolon-chalnicks*, or heads of tables, clerks presiding over the writers or copyists, are in three sections. Two receive 900, two 824, and four 750 roubles. The adjuncts or assistants do not all receive the same amount of pay, but salaries ranging to from 300 to 674 roubles a year. The commissioner has, in addition to his salary, an allowance of 400 roubles a year for travelling expenses, and the assistants have one of 300 roubles.

'Notwithstanding the small number of the officials in the Forest Administration, and the small salaries paid to them, and the salary of the director not being included in the account, the expenses have amounted to the following sums :—

In 1868, subsequent to the reduction in the number of
    officials, -   -   -   -   -   -   -   27,210 roubles.
1867, previous to that reduction, -   -   -   -   31,245 ,,
1866, - - - - - - - - - 31,875 ,,
1865, - - - - - - - - - 31,857 ,,

    Giving as the average of the four years, - 30,557 roubles.

'Each forest administration is divided into four sub-
divisions, with officials bearing corresponding titles; one
of these is the under forester, on whom is devolved the
charge of sales and fellings. Of these there are in all
192.

'When the forest is not large, and is unimportant, they
employ as under foresters warders, who receive the designation of controul guards, and a somewhat higher pay than
the other warders, discharging as they do all the functions
of under foresters. Of these there are 22 in all. In
accordance with this, the whole of the Crown forests are
divided into 214 sub-divisions, to each of which an under
forester is appointed, with an average of 2·1 desatins
for each.

'Of forest warders there are 92 mounted guards; but
the principal ward and responsibility rests on the foot
guards, of whom there are 1072, giving to each a district
of 660 desatins, which, considering the populousness of
many places, is a ward of very considerable extent. In
view of this, in the more dangerous wards there are also
under guards, of whom at present there are 345.

'The local forest administrations in the different Governments are classified as those of forests, and those of sub-
forests. The number of the whole is 63, in which are
included those of the latter category, the distinction
depending on the size of the forests. The forest administrations are committed to chief foresters, or *Oberforst-
Meisters*, each of whom has an assistant supposed to act as
his clerk, with the small salary of from 90 to 100 roubles.
Generally young men, who have newly entered the forest
service on leaving the school of forestry, hold such appointments. The sub-forests are under the charge of *Forst-*

Meisters, sometimes called Sub-foresters, [a designation not to be confounded with that of under forester, another class of officials of whom mention will be made. They, like the ecclesiastical officials designated bishops and arch-bishops in Russia, are equal in power in their own sphere without any subordination to authority of the *Oberforst-Meisters*, and take their inferior rank from the compara-tively lesser importance and magnitude of their charge alone.]

'If we divide the whole area of 710,000 desatins equally amongst the whole of the warders employed, it gives us not less than 500 desatins, which is no small area for a man to guard in those forest estates, which are surrounded by villages in close proximity to them. And notwith-standing thefts of all kinds being carried on by the popu-lation of the locality, and the difficulty of keeping watch and ward, the warders are badly paid. The head warders, for instance, receive an annual salary of from 22 to 45 roubles, and there are some in the service who are only allowed a rouble and a half—[at the present rate of exchange, *three shillings* per annum !]

'The assistant warders receive from 1 to 18' roubles, but with 7½ desatins of land and a free house. With the great scarcity of provisions in Poland, and the poor soil in these small allòttments, the condition of such men was very bad, so bad that it is difficult not to believe that many of them do abuse their official power to make up their insufficient salary.

'Unhappily the continued abnormal condition of the country since 1865 and 1866 prevents us from estimating at their true worth the services of these warders. And although when astounded by the great loss occasioned by theft we cannot attribute this entirely, directly or indirectly, to the warders, it is a fact that on the guarding of the forests being entrusted to the troops the thefts were at once diminished *nine-tenths*.

In 1865 the estimated value of the wood
stolen was   -   -   -   -   -   99,840 roubles.
And in 1866 it was   -   -   -   -   142,850   ,,
But in 1867 it fell to   -   -   -   -   15,902   ,,

The total loss in these three years was -   258,592   ,,

And M. Bitney remarks :—'Considering the very small remuneration given to all of the warders for their work, it would be useless to go into minute estimates of the worth of warders receiving salaries of one and one and a-half roubles per annum.

'For these warders, both foot and horse, they choose generally men who can read and write, who are living in the locality, and soldiers who have left the service with medals for good conduct and attention to duty. These warders may be promoted to be controul warders ; but for controul warders to rise higher it is difficult, as for higher duties a special training and an acquaintance with forest science is required.

'Under every forest administration there are many employés called *Sajeennicks ;* but at present they have no regular employment. The foresters use them as messengers, and they help the warders in watching the forests. Formerly, when firewood was cut up and sold in the Government forests, it was the work of these men to pile the billets in square *sajeens* or fathoms, and thence came the designation. At present this duty has to be performed only in the forests conceded to mines, in which the wood is felled and prepared by Government for use in these. The *Sajeennicks* received no salary, but were allowed so much for each fathom of wood which they piled, and since this practice was discontinued, even this they have lost. This would have no effect on their relations with Government if the forest administration had not placed them in the Crown forests as persons included in the staff of officials employed in the management of the forests. Of course they have no pay, they are of very little use, and it is suspected that in an underhand way they encourage thieving. The number of them in the forests is about 340.

'To all persons employed officially in the management of the forests, from the director-in-chief to the assistant warders and *Sajecnnicks*, there are allotted indiscriminately 7½ desatins of land, as supplying a means of ekeing out their pay. There are thus allotted in all 17,000 desatins. But as these allottments are in the interiors of the forests, and a great many of them are unproductive sands, requiring much labour to make them productive, it follows, to the disadvantage of the service, that the forest warder, receiving but little pay, must necessarily lose or waste, in the culture of these fields, much time which might be spent more profitably in the forest or service. Taking into account the topographical position, one would think that it would be better to give them higher pay, and otherwise provide for their comfort, than to make them engage in agriculture for which the old soldier is very unfit, and cannot get together the implements and other requisites. As it is desired that these people should live within the precincts of the forest, houses have been built for them within the forest itself, but the condition of these houses in many places is very unsatisfactory. In visiting different forest administrations I have frequently seen half-ruined huts which had been assigned to warders, and even to sub-foresters. The houses of the foresters were seemingly in better condition; but many of them also were in need of important repairs.

'All these houses and huts are insured against fire, which entails an expenditure of 19,000 roubles a year. In contemplation of the possibility of all the Crown forests of the kingdom being disposed of by sale, financial considerations make the supreme administrations less careful than would otherwise be the case about keeping all the houses in thorough repair and good condition.

'In the mining districts the forest service is more laborious than it is on the other Crown forests, and on this account the pay of foresters and sub-foresters in these is somewhat higher. The difference is as follows :—

	In forests at large.	In mine forests.
Foresters' annual salary,	525 roubles.	600 roubles.
Assistant or clerk,	90 ,,	100 ,,
Sub-Forester,	180-225 ,,	250 ,,

'The total expense of the forest management was—

In 1865,	-	-	-	119,143 roubles.
1861,	-	-	-	122,755 ,,
1867,	-	-	-	120,022 ,,

Or on an average, 120,640 roubles a year.

'The activity and work of the Polish forester is, in general, much below that of the Russian corresponding official. The reason is two-fold : (1) The excessive centralisation of the forest administration ; and (2) connected therewith the bureaucratic character of the administration. 'Through this the foresters are almost deprived of responsibility. They must obtain authority for the most trifling act, and even in cases in which the forester is allowed to act on his own responsibility, the writings or documents extend to ridiculous dimensions. Count Berg, on becoming acquainted with the administration and actual management of forests in Poland, remarked that he had never even dreamed of meeting with such voluminous and innumerable documents as were the communications of Polish foresters. From this arise great complications and confusion. For example, there occurs a case of theft for which a fine is imposed—in this case all the particulars must be detailed in writing twelve times; and should the accused resist payment, so that extreme measures must be adopted, the correspondence is endless.

'All of the Crown forests, with the exception of those taken over from the clergy of the Church of Rome, are managed systematically, and exploited in accordance with the advanced forest science of the day. This arrangement was effected within the period included in the years 1824 and 1830, so that the greater number have, or will in a few years have passed through their first prescribed course of treatment. [By this, I presume, may be meant a complete revolution, but it does not appear whether that or the ordinary course of felling be what is meant; and this is referred to as supplying the means of forming an opinion in regard to the practicability and expediency of perpetuating the system initiated.]

'The management adopted is substantially the same as that followed in many forests in Russia—a preparation of the forest for regular treatment in accordance with the advanced forest science of the day, by the preparation of accurate diagrams, dividing the forests into compartments, determining the area of fellings for specified periods, and portioning the whole area of the forest so as to secure the requisite products in accordance with an economical use of the capabilities of the forest; and by projecting the exploitation and other operations to be followed out in the course of the decade, including the opening up of the forest by felled strips to give access to all of the compartments required, and making these approximately of equal areas; and by prescribing the operations to be undertaken according as the trees may be young, or middle-aged, or mature, or consisting of a mixture of trees belonging to any two of these categories in any variety of proportions.

'Assuming the operations to embrace a period of thirty years, if the mature section allotted for exploitation in any decade do equal one-third of the whole area, the clearing is confined to the section, and the deficient produce is obtained from other parts; if it exceed the third of the whole area, it is cleared so far as a sale can be assured for the produce.

'It lies with the taxator to prescribe the time and place of all the fellings, and so to combine these as to prepare for the whole forest being brought into such uniformity in its several parts as is most favourable for the carrying out of the most approved forest management of the day, but doing this with the least possible sacrifice of advantages,

—though unhesitatingly risking this if necessary for the execution of the general plan, and to secure plots presenting a regular gradation in age.

'Two copies of his scheme, with charts and specifications, must be prepared, one for the Minister of Finance and the other for the forester; and with the former must be presented a report of the forester's reasons for adopting any measure, or for not adopting any other which may seem naturally to suggest itself for adoption. And the forester can only proceed to carry out the measures proposed when they have received the sanction of the Minister of Finance.

'In subordination to what general scheme of operations it may be deemed desirable, provision must be made for the supply of local requirements. As every section of the forest, of which there are many, is distinct, and surrounded by villages, an opportunity must be supplied to them of purchasing such wood as they require, that they may not be left without this, or be placed under temptation to steal.

'Where the fellings are well conducted, and the soil is adapted for the natural germination of the seed, as, for example, is the case in the Kshepitsk forest, in the Government of Petrokoff, the forest district readily takes the appearance and condition of a regulated forest. But unhappily the fellings throughout the Government forests do not all fulfil these conditions, and the reproduction of the forests felled proceeds very slowly, and in some cases does not take place at all, of which a striking illustration is supplied by the Raigrod forest, in the Government of Lonija, where there are twelve cleared fellings, where there are no signs of reproduction, though a long time has elapsed since they were cleared—in the case of the first felled, fourteen years--and they have still growing on them *baliveaux*, or seed-bearing trees, left in the first felling for the production of seed.

'The soil in these clearings, originally poor, becomes further impoverished by the long continued action of the light promoting the decomposition of the humus ; and from the continuance of this influence a natural reproduction of the forest there can scarcely be expected.

'This system of making a complete clearance, leaving only *baliveaux* for seed, is very much followed throughout the kingdom, and notwithstanding the discouraging results which have been witnessed, occasioning in many cases considerable loss, they are still continued even in the most valuable forests, where, in view of the effects on the revenue, and of the natural hindrances to reproduction, it might have been advantageous to have sacrificed a portion of the proceeds to effect without loss of time an artificial reproduction.'

M. Bitney mentions that in the Government of Plotsk, in a forest growing on very poor sandy soil, the natural sowing of fellings so cleared is proved by many cases to proceed very slowly, requiring from ten to thirteen years, and even then it is very irregularly done. Yet while this is one of the most pecuniarily productive of the forests, yielding annually 1·75 roubles of nett returns for each desatin, the fellings are conducted in the manner, and with the result stated, that of a tardy and poor reproduction of the forest through the false economy of leaving the space to the chances incidental here to natural sowings.

'With regard to the sections of the forest which are in the second and third stages of reproduction and growth, their condition is in general most satisfactory, especially when we take into account that these forests were in the locality where, for about two years, the fighting chiefly took place, and for some time after that they were left open to depredations, which were only repressed by the forests being placed under the guardianship of the soldiers ; but we have seen elsewhere that forests have suffered in this way chiefly on their outer verge, and that to such an extent that there were left there only bare poles, while the

E

centres of the forests remained in a great measure free from
depredation. And if one meets with a few mature or
middle-aged woods, as in Petrokoff, Vlotslaff, and Kam-
pinoss forests, in which are indications of the destruction
of trees, and in which are but a few trees of mature or
middle age, this is more attributable to former long continued mismanagement than to depredations.

'The satisfactory state of the forests, with this exception,
which is surpassing expectation, may be attributable in a
great measure to the system of management adopted,
bringing all as soon as practicable into a regular succession
of growth, and after that subjecting them to a regular
system of exploitation. Through this judicious course,
and the carrying out strictly of the taxator or surveyor's
specific directions, from the first introduction of the new
regimé, the good results have been obtained.

'The number of woods in which the fellings are somewhat objectionable has been greatly diminished; and they
are only to be met with in forests from which it is not
easy to effect sales, and in which the course of management could not, without pecuniary sacrifice, be carried out
within the prescribed time. The whole might be pro-
nounced satisfactory if the restoration or renewal of the
woods had not been so imperfect in some cases in consequence of the character of the soil in the locality of the
forest.

'The sales in the regulated forests are made in accordance with the annual estimates sanctioned by the
Minister of Finance. The wood sold was formerly
delivered to the purchasers—some of it cut, as for example
firewood, and some of it as standing timber, to be felled by
himself; this was done chiefly with wood for buildings,
and under an obligation to have it felled and removed
within a specified time. Latterly, all is sold standing, so
that the Forest Administration does not need to occupy
itself with the felling of the trees, excepting in the mining
districts; and this is done, although in some of the forests,

such as those of Plotsk, Lajnoss, Kshepitsk, and others, it would have been commercially advantageous to have sold the wood cut.

'The sale of wood from Crown forests was formerly made at a fixed price. Sales by auction were only made in exceptional cases, but within the last two or three years this has become the principal mode of sale. And thus the income from some forests has been increased 20 per cent.; but there are not a few forests the income from which has been diminished.

'This may be attributed to the circumstance that the general order issued to the local Forest Administration by the Central Administration was not given in detailed directions; and the administrators of forest estates, which could only command a local sale to the rural population around, had to sell much of the produce by auction to greater buyers, as they had more produce than supplied the local requirements, and it was of considerable value. The Central Administration having approved of the subdivision of sales, this was tried, but the sales were held in but a few places, and were made solely as an experiment, and the small consumers were not accustomed to buy standing trees direct from the Crown, but to get their wood from middlemen, who were Jews with capital at command. The good effects of auction sales have not yet developed themselves.

'The accessory preparations from forest products have only been manufactured to a limited extent in Poland. The *fabricks* met with are principally of tar, and are on the model of those used a hundred years ago—small and inconvenient, and when there are any of a larger and more comprehensive character, as for example, about Brok and Vishkoff, they are pointed out as something remarkable. The making of birch tar and lampblack is carried on to a limited extent. Bark for use in tanning is collected more in private forests than in those of the Government; but from the long time required for its production, and the supervision of the shrubs which is necessary, the quality is poor, and the price low.

'The depasturing of cattle in the forests, which is very injurious to these, though prohibited, is practised almost everywhere without permission. It was kept in check before the insurrection by raising the charge for grazing in the woods. It is to be hoped that this abuse will be rectified, so far as is compatible with rights lawfully acquired, so soon as the forests shall pass out of the transition condition in which they have been for the last five years.

'The collection of moss, fallen leaves, and twigs, is carried on chiefly through such acquired right; but the right is very limited, so the regulation of this presents fewer difficulties than the limitation or prevention of the feeding of cattle.

'The leased woods are confined to a very few forests; the revenue derived from them is trifling, although it might easily be increased. The sales by auction are only of fellings for one year.'

'The following is a statement of the cubic forest income and expenditure for the last three years :—

EXPENDITURE.

	1865.	1866.	1867.
a. For local forest management, insurance, houses, and office expenses of the Forest Administration, - - -	125,284	129,157	127,464
b. Forest improvement and culture, - -	4,621	8,488	5,103
c. Forest section in offices of the several Governments, - - - - -	15,390	15,390	18,390
d. Salaries of officials in the Central Forest Administration, - - - -	31,875	31,875	31,245
Total,	179,170	184,910	182,372

INCOME.

	1865.	1866.	1867.
a. Received from sale of wood, payments due but not paid, fines, and *baliveaux*,	465,499	498,255	423,306
b. Accessory produce sales and arrears, -	18,268	18,975	15,345
Total,	483,767	517,230	483,651

Giving an average annual income for these
 years of . . . .   497,883 roubles.
And an average annual expenditure of .   182,150 „

Nett proceeds,   297,733 „

' This is equal to 42 kopecs of clear revenue from every
desatin of forest land, marshes, and lands, assigned to the
employés, all included. Comparing this result with the
results reported in the annual report of the Forest Administration in Russia for 1866, it appears that from the
forests of Poland there is obtained a nett revenue equal
to that derived from the forests in the Governments of
Moscow and Bessarabia. It is below what is obtained
from the forests in the Governments of Kaluga, of Tula,
of Voronetz, and Poltava; but all of the other 43 Governments of Russia yield a less revenue than does the kingdom of Poland.'

The statements made are a free translation, with omis-
sions, from the report by M. Bitney; and to protect him
from responsibility for errors in translation the narrative
form has in some places been made use of.

Of trees indigenous to Poland it was stated in a letter
addressed by Mr Hove, a native of Poland, to Sir John
Sinclair, when he was compiling his Report on the Agricultural State and Political Condition of Scotland :—

' In Poland there are three sorts of oaks, the *quercus
robur*, or the common oak; the cerris; and another sort,
with which I have not met anywhere else on my travels in
Europe, except on the river Bug; this is the sort which
supplies the English navy with their crown planks. This
tree has hardly any lateral branches in its infant state,
which are so common to all the other known sorts. After
having raised itself from the acorn to the height of seven
feet, it assumes a diagonal form, or position, and the tops
of such trees in the plantations are quite entangled with
each other; but, on arriving at the age of fifteen or sixteen years, they acquire a height of from twenty-four to
thirty feet, begin to form a crown, gradually erect them-

selves, and become majestic and stately trees. The leaves
of this tree are much narrower, longer, and more deeply
cut than in the *robur*; the bark is perfectly smooth, and
the acorn long and pointed. On my leaving the district
of Belsk, where they grew, five years ago, but few of these
trees remained, as the Jews, who are the renters and
fellers of timber, had cut them down indiscriminately,
with a view to immediate profit. These rich and immense
forests, which skirted the river Bug, where I
botanised sixteen years ago, are now no more, there
remaining only a few trees very thinly scattered, which
owe their existence only to the circumstance of their
being in situations far distant from the river. I procured
a considerable quantity of the acorns on my leaving
Poland, with the view of enriching this country; but,
having sent my collections *via* Dantzig, where the French
arrived shortly after, I am at this moment ignorant of the
manner in which they have been disposed of. Two
hundred bushels of acorns of this valuable species would
certainly be a great acquisition, if not a real source of
riches, to this country; they would answer for hedge-
planting perfectly well.

'The Swirk is another tree that would be of great value
to this country; it is a species of fir that is peculiar to
the mountains of Pokutia, or mountains of Penitence, to
which Ovid was exiled. The height and bulk of this tree
is incredible; and it is not very nice in regard to soil, as
it grows in the most rocky and inclement situations on
the mountains. The white ash of the Palatinate of Belsk,
and a sort of maple, are trees that would also be of great
value in England; they both grow to an immense height.
The Polish king, John Sobiesky, was so struck with the
size and beauty of these trees, that he built himself a
residence in the neighbourhood of the forests where they
grew, and formed a large town, which is still in existence,
to which he gave the name of Jaworow; Jawor denoting,
in the Polish language, this species of acer. The black

birch, in the same Palatinate, in the circle of Mosciska,* is a new and unknown species. The wood of this tree is more solid than in any other of this genus, on which account the wheelwrights and millwrights give it the preference. The quality of this wood is in such repute that it is sent to Warsaw, and all over Prussia, for their use.'

One of the forest products of Poland, which at one time was in great demand, is honey. Of Polish honey there are three varieties. In regard to these the following information was supplied by Mr Hove to Sir John Sinclair :—

'Honey, another rural production, respecting which you are anxious to procure information, is in Poland divided into three classes, namely Lipiec, Leszny, and Stepowey prasznymird.

'Lipiec is gathered by the bees from the lime-tree alone, and is considered on the Continent most valuable, not only for the superiority of its flavour, but also for the estimation in which it is held as an arcanum, in pulmonary complaints, containing very little wax, and being consequently less heating in its nature; it is as white as milk, and is only to be met with in the lime forests in the neighbourhood of the town of Kovno, in Lithuana. The great demand for this honey occasions it to bear a high price, in so much that I have known a small barrel, containing hardly one pound weight, sell for two ducats on the spot. This species of the lime-tree is peculiar to the province of Lithuania; it is quite different from all the rest of the genus Tila that I met with in my researches in Poland, and is called Kamienna Lipa, or stone-lime. It is a stately tree, and grows in the shape of a pyramid; the leaves are very small, and the twigs uncommonly slender; it flowers in the months of June and July; the flowers are very minute, but more abun-

---

* An estate belonging to Count Palatine de Cetner.

dant than in any other species.   In the Polish language,
the month of June, which is also called Lipiec, derives its
name from the flowering of this tree, as the month of
July derives its name from the Cocus polonicus, called by
the Poles Czerwiee, in which month the ova are gathered.
The inhabitants have no regular bee-hives about Kovno;
every peasant who is desirous of rearing bees goes into
a forest in a district belonging to his master, without
even his leave, makes a longitudinal hollow aperture or
apertures in the trunk of a tree, or in the collateral
branches, about three feet in length, one foot broad, and
about a foot deep, where he deposits his bees, leaves them
some food, but pays very little further attention to them
until late in the autumn; when, after cutting out some of
their honey, and leaving some for their maintenance,
he secures the aperture properly with clay and straw
against the frost and inclemency of the approaching
season; these tenements (if they may be so called), with
their inhabitants, and the produce of their labour, have
then become his indisputable property; he may sell them,
transfer them; in short, he may do whatever he pleases
with them; and never is it heard that any depredation is
committed on them (that by the bear excepted.) In Poland
the laws are particularly severe against robbers or destroyers
of this property, punishing the offender, when detected,
by cutting out the navel, and drawing out his intestines
round and round the very tree which he has robbed.
Such thefts have happened, but not in my memory.

' The following spring the proprietor goes again to the
forest, examines the bees, and ascertains whether there is
sufficient food left till they are able to maintain themselves; should there not be a sufficient quantity he
deposits with them as much as he judges necessary till
the spring-blossom appears.   If he observes that his
stock has not decreased by mortality he makes more of
these apertures in the lateral branches, or in the trunk
of the tree, that in case the bees should swarm in his
absence they may have a ready asylum.   In the autumn

he visits them again, carries the June and July work away with him, which is the Lipiec, and leaves only that part for their food which was gathered by them before, and after the flowering of the lime-tree. I have not the least doubt, that if this species of the lime-tree were introduced, and attention paid to them, honey equally excellent and valuable might be produced in this country. The mead made from this honey is excellent, it is sold at Kovno, Grodno, and Vilna, at the rate of £8 the dozen.

'The next class of honey, which is inferior in a great degree to the Lipiec, being only used for the common mead, is that of the pine forests, the inhabitants of which make apertures in the pine-trees similar to those near Kovno, and pay the same attention in regard to the security of the bees, and their maintenance. The wax also is much inferior in quality; it requires more trouble in the bleaching, and is only made use of in the churches.

' The third class of honey is the Stepowey, or the honey from the plains, where there is an abundance of perennial plants, and hardly any wood. The province of Ukraine produces the very best, and also the very best wax. In that province the peasants pay particular attention to this branch of economy, as it is the only resource they have to enable them to defray the taxes levied in Russia; and they consider the produce of bees equal to ready money: wheat, and other species of corn, being so very fluctuating in price, some years it being of so little value that it is not worth the peasant's trouble to gather it in. This has happened in the Ukraine four times in twelve years. But for honey and wax there is always a great demand all over Europe, and even in Turkey. Some of the peasants have from four to five hundred Ule, or logs of wood, in their bee gardens, which are called Pasieka, or bee-hives; these logs are about six feet high, commonly of birch wood (the bees prefer the birch to any other wood), hollowed out in the middle for about five feet, several lamina of thin boards are nailed before the aperture, and but a small

hole left in the middle of one of these for the entrance of
the bees. As the bees are often capricious at the beginning of their work, frequently commencing it at the front
rather than the back, the peasants cover the aperture
with a number of these thin boards, instead of one entire
board, for fear of disturbing them, should they have begun
their work at the front. It may appear extraordinary,
but it is nevertheless true, that in some favourable seasons
this aperture of five feet in length, and a foot wide, is full
before August, and the peasants are obliged to take the
produce long before the usual time, with the view of
giving room to the bees to continue their work, so favourable is the harvest some years.

' The bee-gardens are chosen in the plains where the
perennial plants are most abundant, that the bees may
have but little way to carry home the produce of their
labour ; they are of a circular form, about 150 yards in
diameter, enclosed with a fence of reeds or brush-wood,
and a thatching over them of about five feet for protection, and to keep out the rain and snow ; this is supported by poles from the inside, and a bank of earth is
also thrown up to keep the snow from penetrating there in
the winter. In the middle a few fruit trees are planted
to break the wind, as also hawthorns, and other under-
wood, round the enclosure, with the same view. The
hives are planted under cover, in the inside, round the
fence, and in the winter they are well secured with straw from
the frost. The plants for which the bees have a preference are the *Thymus serpyllum, Hyssopus officinalis,
Cerinthe maculata,* and the *Pollygonum fagopyrnm.*

' The process of brewing mead in Poland is very simple :
the proportion is three parts of water to one of honey,
and 50 lb. of mild hops to 160 gallons, which is called a
Waar, or a brewing. When the water is boiling, both the
honey and hops are thrown into it, and it is kept stirring
until it becomes milk warm ; it is then put into a large
cask, and allowed to ferment for a few days ; it is then

drawn off into another cask, wherein there has been aqua-
vita or whisky, bunged quite close, and afterwards taken
to the cellars, which in this country are excellent and cool.
This mead becomes good in three years time, and by
keeping, it improves, like many sorts of wine. The
mead for immediate drink is made from malt, hops, and
honey, in the same proportion, and undergoes a similar
process. In Hungary it is usual to put ginger in mead.
There are other sorts of mead in Poland, as Wisniak.
Dereniak, Malinaik; they are made of honey, wild
cherries, berries of the *Cornus mascula*, and raspberries ;
they all undergo the same process, and are most excellent
and wholesome after a few years keeping. I never saw a
gouty man in those provinces where mead is in common
use. The Lipiec is made in the same way, but it contains
the honey and pure water only. The honey gathered by
the bees from the *Azalea pontica*, at Oczakoff, and in
Potesia in Poland, is of an intoxicating nature; it produces nausea, and is used only for medical purposes,
chiefly in rheumatism, scrophula, and eruptions of the
skin, in which complaints it has been attended with great
success. In a disease among the hogs called *wengry* (a
sort of plague among these animals), a decoction of the
leaves and buds of this plant is given, with the greatest
effect, and produces almost instantaneous relief. This
disease attacks the hogs with a swelling of their throat,
and terminates in large hard knots, not unlike the plague,
on which the decoction acts as a digestive, abates the
fever directly in the first stage, and suppurates the knots.
It is used in Turkey, I understand, with the same view
in the plague. Tournefort makes mention, in his travels,
of this honey.'

CHAPTER IV.

A METHOD of exploitation so complicate as that which has been described can only be administered by educated men, specially instructed and trained for the work.

The administration of the Crown forests in Poland is subject to the Imperial Minister of Finance. In a communication I had from Count Ostenstacken, personal attaché to Field-Marshal Berg, Viceroy of Poland, I was informed that the private proprietors of forests within the kingdom act very recklessly in disposing of these and in deforesting the country.

Their doing so renders it the more desirable that the Crown forests should be conserved and properly managed, if not also extended. And, in view of this, provision has been made for the education, instruction, and training of candidates for employment on the forest service.

Besides the Imperial Schools of Forestry in the vicinity of St. Petersburg, and in the vicinity of Moscow, and others in Eastern and Central Russia, in Western Russia a Chair of Forestry has been established in the Institutes of Kalazin, Vologda, Lipetsk, in Marymont in Poland, and at Mittau. But the most important arrangements for the study of forest science and forest economy by forest officials in Poland are at Novoi Alexandria, or New Alexandria.

Of these arrangements the following details are given from the *Ustaff*, or Code of Regulations for the Institute of Agriculture and Forestry at that place, approved by the Emperor 8th June, O.S., 1869, and still in force.

' The institute is ranked as a college of the first class, with two sections—one devoted to the study of rural economy and agriculture, the other to the study of forest science and forestry, with a farm, a forest, and an extensive domain attached to it, the whole being placed under the Minister of Public Instruction at Warsaw.

'The staff of officials includes a director, an inspector, five professors, eight tutors, and three teachers, a laboratory superintendent, a mechanic, a foreman of the workshop, a land steward or manager of the estate, a gardener and assistant, a surgeon, a secretary, a book-keeper, and a superintendent of buildings.

' The director, nominated by the Minister of Instruction, is at the head of the Institute, having the direction of everything connected with it. The inspector, similarly appointed, has corresponding duties. They, together with the professors, constitute the council of management. No professor can hold two chairs, and any of them after twenty-five years may be again and again re-appointed for successive terms of five years each. The laboratory superintendent has charge of all the laboratories of the Institute. The mechanic has charge of all the implements.

' At meetings of council, under the presidency of the director, all questions are decided by vote, and in certain cases the vote may be taken by ballot.

'A board of management, consisting of the director, inspector, and two professors, has the charge of expenditures to the amount of 300 roubles, say £37 10s, to be sanctioned by the director; the expenditure of sums between 1000 and 5000 roubles, say £125 to £625, require the sanction of the council, and the expenditure of sums above this amount that of the Ministry. These meetings are held weekly.

' The general instruction embraces geometry and practical mechanics, geodesy, plan drawing and land surveying, physics, meteorology and climatology, chemistry, zoology, botany, mineralogy, agricultural technology, soil, cubic increase of trees by growth, architecture, political economy, and statistics of Russia; laws of land tenure and more particularly of the tenure of land in Poland; book-keeping, drawing and preparation of diagrams and plans.

'The special subjects of study in the agricultural section are analytical and agricultural · chemistry, agricultural mechanics, breeding and rearing of cattle, sheep, poultry, and bees; sericulture and pisciculture, and veterinary science and practice.

'In the department of forestry the special subjects of study are forest botany and entomology; forest management and taxation, or calculations of the cubic contents, prospective growth, pecuniary value, &c., of the trees; and forest statistics of Russia and of Poland.

'The instruction is communicated in the Russian language. Each professor and tutor is required to give six lectures a week, and teachers to spend twelve hours a week in class duties. The professors may be required to lecture in both departments.

'The session extends from the 20th of August to the 18th June, O.S.

'All books requisite for study are in the library.

'Students are required to be seventeen years of age, to have attended a high school, and satisfactorily passed the prescribed examination. They wear a uniform, are required to maintain a becoming behaviour, and to be obedient and respectful to their teachers. If any be retained for two years in any class, and be still unable to pass, they are dismissed. Medals, &c., are given to students in recognition of merit.

'The rights of the institution as a first-class college, and the rights of the professors as officers of the Imperial civil service, are formally secured.

'The salary of the director is 3,500 roubles, say £450, with noble rank of the fifth class and civil service pension. The salaries of the inspector and professors are 2000 roubles, say £250,* with noble rank of the sixth class, and corresponding pension.'

---

* The rate of exchange varies. These calculations were correct at the time I received the information. At this time (1885), according to the present rate of exchange, they would be respectively £350 and £250.

I spent in St. Peterburg the summer of 1878, and I was invited to attend a congress of Russian foresters, to be held in Warsaw about the time of my return to England; but a similar congress of German foresters was held about the same time in Dresden, and 1 found it more expedient to return by Saxony.

The correspondent of the *St. Petersburg Exchange Gazette,* writing from Warsaw under date of 22nd August, 1878, O.S., gave the following account of the foresters' conference held there :—
' The details and minutes of the conference are known to you from the reports issued. I wish to communicate the outside or public influence of the conference on the press and on the Polish nation. It is a question whether any other Russian congress, or conference of similar character and aim, has ever produced so great an effect on the public mind. The intelligent classes of the community and the public press exhibited a lively interest in this, the fourth Russian conference of foresters and landlords. The specialists and the owners of forests who had arrived in Warsaw were the heroes of the day, and the questions discussed attracted the liveliest notice of the public. The reports of the conference appeared in print daily; even the illustrated papers contained views of the "Salon" during the sitting of the congress. The vital question of forestry was worthily discussed by the Polish press, and the economical importance of forestry was energetically advocated. The ground was preliminarily prepared by free invitations to attend the congress, and by timely issues of pamphlets and papers on the subject, the object of which was to acquaint the public with the objects of the congress. The intelligence of society in general was adequate to the occasion. Thanks to this " coaching," the public were prepared to give the guests a hearty welcome; and many specialists and owners of land arrived from all parts of Poland. Three hundred members attended the first sitting. But the great attraction was the permission

to discuss in Polish as well as in Russian. Many of the
Polish landowners could not have spoken on scientific
subjects in Russian, and therefore it was a very wise step
on the part of the committee of the congress to permit
the use of the two languages. To the honour of the
committee be it said that it withheld itself from bureaucratic views, and having set aside out-of-place principles,
permitted discussion in Polish, and they thereby enlisted
the active sympathy of many who would otherwise have
been but dumb spectators. Measures of this kind rendered the conference very popular. Russian members
became the "very dear guests" of Polish members, an
enthusiasm such as is seldom witnessed was engendered,
and the weighty subjects were most warmly discussed.
According to accounts of old residents of Warsaw, such
heartiness in a "public important topic" has seldom been
shown. The whole finished up with a "final supper in the
Swiss valley." This entertainment was characterised by
great warmth, and expressions of warm sympathy on the
part of all assembled.'

In reply to an enquiry which I addressed to a friend,
a Pole, in regard to works on forestry and botany in
Polish, he supplied me with the following list:—
  Alexander Potujanski.—Opisanie lasów Królestwa Pol-
skiego i Gubernij Zachodnich Cesarstwa Rossyjskiego pod
wzgledem historycznym, statystycznym i gospodarczym.—
Warszawa, 1846. 4 tomy. (Description of the forests of
the kingdom of Poland and Western Governments of the
Empire of Russia. Warsaw. 4 volumes.)
  J. Waga.—Flora Polonica—Varsoviae, 1848 an., and
Stirpes rariores Bucovinae — Stanislawow 1853 an.
(Bucovina at present belongs to Austria.)
  Flora Cracoviensis, 1859.
  Kluk Krzysztof.—Dykcysnarz roslinny. (Dictionary of
plants.)
  Jundzitt Stanislaw Bonifacy.—Opisanie róslin litew-
skich podlug ukladu Linneusza. 1811. (Description of

the plants of Lithuania according to the system of Linnæus.
1811.)

Jundzitt Józef,—Opisanie róslin w Litwie, na Wolyniu,
Podolu, i Ukrainie, drika rosnacych jako i oswojonych.
Wilno, 1830. (Description of the plants in Lithuania,
Volynia, Podolia, and the Ukraine.)

CHAPTER V.

POLISH HISTORY.

WHILE it is forestry to which my attention has been chiefly given in my studies of Poland, with such a history as that country has, it is natural that this should receive some attention from the student of forest science.

According to the Abbé des Fontaines, in his *History of the Revolutions of Poland from the foundation of that Monarchy to the death of Augustus II*—a work published early in the last century, in which the author followed Duglossius, a canon of Cracow, who composed a history of Poland in Latin, which, though valuable, has defects which the Abbé des Fontaines in his work endeavoured to remedy, while he also adduced additional details from the works of Thuanus, which are of considerable celebrity :—

' If we may credit their own historians, their first prince was a descendent from Japhet, the son of Noah. They give him the name of Lecht, and declare that he came from Dalmatia. This prince left his throne to his son Wissimir, who founded the city of Dantzig. We discover no traces in history of any actions that were performed by the posterity of these two first kings of Poland ; and it is a void which fiction itself has never attempted to fill up. It only supposes that the nation, after the extinction of the royal family, assembled for the election of new masters. The nobility were on the point of proceeding to this choice, when the people, who had long been harassed with the tyranny of their last kings, demanded an abolition of the regal government, that they might no longer depend on the caprice of one man.

' The great lords, who were allured with the hopes of sharing all the honours of dominion, were easily induced

to comply with the solicititations of the people; in consequence of which they established a republic, the administration whereof was intrusted to twelve palatines; but
the unsteady people were soon dissatisfied with this new
mode of government; an anarchy full of disorder and
confusion inspired them with an aversion to their state of
independence, and a set of enemies, who were always
ready to derive advantages from the troubles of the State,
and the conjunctures of those times, ravaged the provinces
with impunity, and made this nation pay dear for the
liberty they had acquired.

'The eyes of the people were at last opened to their
real interest, and they judged it would be most advantageous to them to have but one master. This consideration induced them to turn their thoughts to the election
of a king; but a choice of this kind was attended with
great difficulties. The state of their affairs required
some martial spirit to repel the invasions of the neighbouring people, as well as to re-conquer the territories
which had been wrested from them by usurpations, and to
re-establish the honour of the nation. It was likewise
necessary for this hero to temper an intrepidity of mind,
with the softness of a prudent charity, in order to gain
upon those dispositions, which had been rendered intract-
able by a long state of independence; and it was also
thought requisite, that the virtue of the future prince
should afford them a sufficient security for the proper use
of the supreme power with which they were disposed
to entrust him.

'Such qualities are seldom united in one man; the
Poles, however, found one of their countrymen who
possessed them in an eminent degree. Grack was the name
of this virtuous person who brought the calamities of
Poland to a happy period. As he was always victorious
in the wars he undertook, and as he likewise guided the
reins of Government with a consummate prudence, he
continuoustly rendered himself dear to his people, and
formidable to his enemies. He built the city of Cracow

on the Weissel; and Bohemia submitted to the laws of so accomplished a prince.*

'Lecht the Second became his successor, the consequence of a crime, for he secretly destroyed Grack, his elder brother, and ascended his father's throne, as well by the choice of the nobility, as by virtue of the right he claimed to the succession. All his subjects submitted to his authority while his crime was undiscovered; but as soon as it happened to be detected, the lords would no longer suffer the assassin of his own brother to sit upon the throne he had usurped, and to hold the reins of government with the hands which were defiled with the blood of their lawful prince. He was chased from the kingdom in a degrading manner, and, according to some authors, he died without leaving any children, detested by his subjects, and troubled by the remorse of his own conscience.

'After the death of the two sons of Grack the First, the Poles were desirous of submitting to the government of his daughter Vanda, a very amiable princess, who was graced with the accomplishments of eloquence, wisdom, and courage to a degree that was altogether uncommon in her sex. She reigned with glory, and in the tranquillity of a profound peace, when a neighbouring prince sent ambassadors to her to treat of a marriage between himself and her, and to denounce war against her dominions if she should happen to reject his offers. Vanda, according to some historians, had rendered herself incapable of the nuptial state by a vow of virginity which she had made to the gods of her country. She therefore prepared for war, assembled her troops, animated them by her presence and discourse, broke the measures of her enemy, opposed his incursions, and constrained him at last to come to a conference. What can be impracticable

* His reign is thought to have been coincident with the beginning of the seventh century. Duglossius declares, Lib. i., *Habet nonnullorum assertio Graecum principem ante incarnationem Christi annis circiter quadringentis regnare apud Polonorum gentem cœpisse.* But if he reigned 400 years before the incarnation, what a void will be opened between his reign and that of Miecflaus the First!

to beauty in conjunction with eloquence? Vanda was soon rendered victorious by the sole aid of her charms; she enchanted the hostile troops in a moment, the commanders refused to combat against so amiable a princess, the soldiers quitted their ranks, the most savage among them were disarmed of all their rage, their chief himself was forsaken by all his troops, he yielded to the impressions of confusion and despair, and plunged his sword into his own breast as a punishment for his temerity.

'The princess was easily induced to pardon the foes she had vanquished in this manner; and as she was satisfied with securing the repose of her subjects, she repaired to Cracow, to receive their applause in that city, where they decreed her the honour of a triumph, for a victory she had acquired by her charms and wit, one in which her soldiers could not pretend to any participation. It is a pity that this princess should have become an enthusiast after this great event; but she considered it as an evidence of the favour of her tutelar gods; and she imagined it was incumbent on her to testify her gratitude by a strange sacrifice, in which she herself was the victim. In view of this, she threw herself into the Weissel.

'A death so tragical as this left the Poles a second time without a master, and they had then an inclination to enjoy the sweets of independency. Though liberty had already proved so fatal to them, they were allured by the pleasures they expected to enjoy by changing their state, and they resumed the republican form of government. Poland was then divided into twelve palatinates, the administration of which was committed to the same number of lords, who were chosen to dispense justice to the people, and to defend them against the enemies of their State. This ancient order still subsists among them, and the Palatines have to this day a power, under their kings, almost equal to that which they enjoyed at the time of their first institution.'

From that time onward their history appears a record

of unceasingly recurrent wars and revolutions, reminding
one of what Beshlam, Methridah, and Tabeel wrote to
Artaxerxes in regard to the Jews*; but unhappily for
the credit of the Poles, details are given which cannot be
gainsaid.

In regard to the introduction of Christianity into
Poland, the Abbé writes :—
'The northern nations had already begun to embrace
Christianity; Sclavonia and Bohemia had, for some time,
renounced the errors of paganism, and Poland was converted by a pious princess. Dabrowka, the daughter of
Boleflaus, Duke of Bohemia, was the person by whose
ministration God accomplished this work. This lady was
determined not to espouse Miecflaus, unless he would
consent to be baptised. The prince caused himself to be
instructed in the truths of her religion ; and when he had
declared himself a Christian he was desirous that his people
should follow his example. With this view he became
their apostle; all the idols were destroyed, and on the
ruins of their altars temples were erected to the true
God.
'Miecflaus, some time after his conversion, sent the
Archbishop of Cracow to Rome, to assure the Pope of
his obedience, and to demand from him a donation of the
regal Crown ; but Benedick VII. gave the preference to
Stephen, Duke of Hungary, by whom he had been also
solicited for the same gift. This conduct of the Pope
either sprung from some prejudice he had entertained
against Miecfluas, or perhaps he may have already heard
of the death of that prince.
'He was succeeded by his son, Boleflaus, who was
advanced to years of maturity when he ascended the
throne. His courage was never abated by any difficulties,
and the severest toils of war constituted his pleasures.
He had the abilities of a chief and a soldier, and knew

---
* Ezra iv. 7-16.

how to command and execute at the same time. He
appeared magnificently in public, and whenever it was
requisite of him to assume the air of a great prince ; his
more private conduct was softened with an air of affability ;
and he saw himself respected and beloved by his
people, whom he treated more as a father than as a
Sovereign. His renown was so great that Otho the Third
came into Poland, as well to offer him his alliance, as to
accomplish a vow he had made to the martyr, St. Alder-
bert, or Albert, Archbishop of Gnesna. The Emperor
was so well satisfied with his reception, and likewise with
the magnificence of Boleflaus, that he thought it incumbent on him to testify his acknowledgments to him by
some honourable return that might correspond with the
treatment he had received in the territories of this
prince. He accordingly crowned him King of Poland,
and gave him the imperial eagle, in a field gules, for his
arms. The two princes afterwards confirmed their new
alliance by the marriage of Rixa, or Riche, daughter of
Godfrey, Count Palatine of the Rhine, and niece of the
Emperor, with Miecflaus, the son of Boleflaus.'

In subsequent times, under the provision for an elective
monarchy, Russia, and France, and England, all of them,
as well as other lands, furnished statesmen nominated as
candidates for the throne—England in the person of Sir
Philip Sydney, France in that of Henry of Vallois.

Of the political constitution of Poland in the early part
of the last century, the Abbé writes:—
' The kingdom of Poland is composed of Poland properly
so called, which is divided into the Upper and Lower
Poland, Royal Prussia, the Grand Duchy of Lithuania, and
the provinces of Mazovia, Polachia, Black Russia, Volhinia,
Podolia, the Ukraine, and some other small provinces.
The Baltic Sea, Samogitia, Livonia, and Muscovy constitute its northern bounds. The dominions of the Russian
monarch and Little Tartary form its frontiers to the east.

It is bounded on the south by Moldavia and the mountains of Krapac; and its western limits are Moravia,
Silesia, and the territories of the Elector of Brandenburg.
It extends 260 leagues in length from west to east, and
comprehends 200 leagues in breadth from south to north.

' Gnesna, which is a city in Great Poland, was formerly
the capital of the kingdom, and is still the metropolitan
city. Its archbishop is the head of the republic during
the continuance of any interregnum, and his power is so
great that he has frequently caused kings themselves to
be dethroned.

' Cracow, which is situated on the Weissel, is now considered as the capital of the kingdom, since the sovereigns
have fixed their residence in that city. The royal orna-
ments are deposited there, and it is now the place where
the kings are crowned.

' The Weissel, the Boristhenes, and the Dneister, are its
principal rivers. Commerce might easily flourish in that
country, since nature has supplied the inhabitants
with every provision capable of facilitating navigation and
traffic with Europe and Asia; but these advantages are
disregarded by them. The gentry are devoted to arms
alone, and the peasants are a race of miserable and unin-
dustrious people, who are crushed under the yoke of their
lords, and have no property which they can call their own.

' Poland is not strengthened by any fortified cities, and
every place is entirely open and free. The Poles consider
castles and fortresses as so many rocks on which their
independency would be wrecked, and as the tyrannical
instruments of some ambitious person, who would be
desirous to load them with chains. Kaminieck, on the frontiers of Moldavia, is not considerable either for its extent
or fortifications; and Dantzig, which is the strongest
city in Poland, is but moderately fortified.

' The king, who is considered as the first magistrate in
the Republic, derives all his authority from that of the
nation, and if he should happen to abuse the power con-
fided to him, and should refuse to conform to the compact

made between the people and himself at his coronation,
he would soon behold a potent confederacy formed against
him for his deposition. He is incapable of making new
laws, raising taxes, contracting alliances, or declaring war
without the ratification of a diet; nor can he even marry
without the permission of the States.   Moreover, the
prince is not authorised to coin money, since this prerogative is peculiar to the Republic. His revenues amount to
no more than a million of livres, but he only defrays the
expense of his table; all other charges are paid by the
republic.

'The senate is composed of the clergy and the nobility,
for the third state is not so much as known in Poland.
The grand marshal, the marshal of the court, the
chancellor, the vice-chancellor, and the treasurer, are the
first senators.   The kingdom of Poland and the Grand
Duchy of Lithuania have alike all these officers.

'The grand marshal is the supreme judge of all disorders which at any time happen in the diets and the
king's household. He imposes silence, and authorises
freedom of speech in the National Assemblies.   He
introduces ambassadors, examines their dispatches, and
assigns to them their apartments. He likewise determines the price of all merchandise whatsoever it may be.

' The marshal of the court, or the deputy-marshal, is his
substitute, and discharges all his functions in his absence.

' The chancellor is intrusted with the seals of the kingdom, and even the sovereign cannot compel him to affix
them to any decrees without the privity and approbation
of the States. All civil affairs, and those which relate to
the king's domain, are brought to his tribunal. He is
charged with the preservation of the laws and the preservation of liberty. He in the diets returns answers to the
ministers of foreign powers; and if he happens to be an
ecclesiastic, he extends his inspection to the secretaries,
the priests, and preachers at court.

' The jurisdiction of the vice-chancellor is exerted only
in the absence of the grand chancellor, but he is in pos-

session of seals as well as the other. These two great posts are held alternately by a spiritual and by a temporal lord.

'The money which belongs to the republic is deposited with the treasurer, who regulates the revenues; and he ought to assist in all contracts made by the king, which have no validity till they have been signed by this officer.

'Those who compose the senate, next in station to these ten prime officers of the kingdom and the Grand Duchy, are the bishops, the palatines, the castellans, and some starosts, who preserve there the rank annexed to the dignity of their bishoprics, palatinates, castellanies, and starosties.

'A palatine commands the troops of the particular province which is consigned to his Government; he is the president of the nobility of his palatinate, and his jurisdiction extends to civil and criminal affairs.

'The castellans are the lieutenants of the palatines; and the starosts, or captains, have much the same rank. But though the palatines generally precede the castellans and the starosts, yet the castellan of Cracow is, by a peculiar privilege, superior to the palatine of that city; and the starost, or captain general, of Samogitia, which is a vassal province of the Republic of Poland, takes place of several Polish and Lithuanian palatines.

'The clergy, who constitute the first order of men in the kingdom, are rich and powerful; they possess more than 200,000 towns, and several considerable cities. The power of the secular clergy, is, however, balanced by that of the monks, who invade the privileges of the common pastors in a thousand instances with impunity, and cause themselves to be dreaded and respected, in consequence of the empire they have assumed over the minds of a credulous people.

'The gentry compose the second order, and they possess dignities and employments, as well in the kingdom as in the Grand Duchy, and in which they never permit strangers or the commonalty to have the least par-

ticipation. They are privileged to elect their kings, and
the senators have involved themselves in the greatest
dangers whenever they have discovered an inclination to
render themselves the masters of the election.

'When the kingdom is threatened with any irruption
the pospolite, or the whole body of the nobility, are armed
and mount their horses. Besides the palatines of each
province, who appear at the head of their respective
nobility, this body is commanded by a general, even when
the king himself is present. The nobility who compose
these troops are very magnificent and courageous; they
are covered in their march with the skins of tigers, leo-
pards, and panthers ; their horses are full of mettle, and
their furniture is very splendid ; but the gentlemen pay
no extraordinary obedience to the orders of their chiefs.
They neglect with impunity to assemble at the place
appointed by the letters of convocation, and when they
happen not to be paid, which is generally the case, they
disband themselves without any previous discharge.
Their march is altogether as irregular, they commit a
thousand disorders in the kingdom, and as there are never
any settlers in the Polish army, and as no care is taken to
erect magazines, they make no scruple to pillage where-
ever they come.

'The peasants are in a slavish subjection to the gentry,
They have no property of their own, and all their acquisitions are made for their masters. They are indispensably
employed in the culture of the earth, and they live in an
absolute state of servitude. They are incapable of engag-
ing in any state of life which would procure them their
freedom, without the permission of their lords, and they
are exposed to all the effects of the ill-disposition of
their tyrants who oppress them with impunity.

'The general diets are usually held either at Warsaw
or at Grodno, in the Lithuanian palatinate of Troki.
These are always preceded by dietines, or particular
assemblies of palatinates, in which they choose their nuncios, or deputies for the general assembly; and their

several instructions, with the demands they are to form, in the name of the province, are regulated in these dietines.

'The king convokes the diet by dispatching letters, which are called Universalia, to all the palatinates; and yet the nobility assembled in the reign of John Casimir the second, without the orders of that prince, and the Polish pospolite, marched into the Ukraine, in order to subdue the Cossacks, contrary to his desire.

'The nuncios, who are elected in the dietines, meet at the place specified by letters of convocation, and take their seats in the assembly, according to the order and dignity of the palatines which they represent. They afterwards proceed to the election of a marshal of the nuncios, who is alternatively chosen out of the lords of Great and Little Poland and Lithuania.

'This officer has a very extraordinary power in the diet. No member whatever can speak without his permission, and he is empowered to impose silence on whom he pleases. He is the organ of the nobility, and transmits to the senate and the sovereign all complaints against exorbitant proceedings, abuses in Government, and injurious treatment of particular persons; he is attentive to the protection and safety of the deputies, whose chief he is constituted by his office, and he punishes all offences that are committed in the assembly.

'The general diet for the election of a king is that wherein strangers are interested the most. When an interregnum has been declared, either in consequence of the death, abdication, or deposition of the king, the primate, who is then the chief of the republic, dispatches his universalia to the several provinces, for a general assembly. A deputation of senators is likewise sent to the army to assist the generals with their counsels, and an exact inventory is taken of the treasure of the Crown. All the tribunals are then discontinued, and every jurisdiction, except those of the marshals, entirely ceases.

'The assembly is at last held near Warsaw, in the open field, which is surrounded with ditches, and covered over with boards. The Poles call it the Szopa or the Colo. On the day fixed for opening the diet, the senators and the nuncios are present at the celebration of a mass of the Holy Ghost in the church of St John in Warsaw, after which they repair to the Colo, and when they have elected a marshal of the nuncios they form a confederation or a treaty, by which the members of the diet take an oath not to separate till they have elected a king, and not to acknowledge any candidate, if he has not been elected by their unanimous approval, nor render to him any act of obedience, till he has sworn to observe the *Pacta Conventa*, and the other laws of the kingdom.

'When this union has been formed the members enquire into the exorbitances that have been committed in the course of the last reign. Though the authority of the prince be limited to very narrow restrictions, and though the jealousy which the nation entertains of all attempts against its independency prompts it to a scrupulous examination of their prince's conduct, yet there are always some points to be complained of, and reformed at the close of every reign, and the interregnum proves a favourable opportunity for the correction of those abuses. The laws are re-established in their original force, and new ones are likewise added; all customs, that are inconsistent with the immunities of the nobility, are rectified. They likewise regulate the administration of the State, and prescribe to their future king the observance of those rules and duties from which he is never permitted to deviate.

'All ambassadors are introduced by the senators, and they address the assembly in Latin. The president answers them in the name of the senate, and the marshal of the nuncios on the part of the nobility.

'It is incumbent on the ministers of the candidates to let their gold glitter as much as possible; they ought to give splendid entertainments, which, besides their pomp, must be carried into debauch; and nothing is more agreeable to the Poles, who are naturally magnificent, than are feasts of this kind. The nobility are captivated in a peculiar manner with the attractions of Hungarian wine, and infallibly declare in favour of the candidate who causes it to flow in the greatest profusion.

'The confederates usually take an oath not to attach themselves to any particular faction, and the ministers are prohibited from continuing at Warsaw, or forming any cabals, but these injunctions are always ill observed. The ambassadors enter upon intrigues even in public, the nobility receive their presents, sell their suffrages with impunity, and render the throne venal, after their own infraction of the first article of the confederation. These mercenary gentlemen usually conduct themselves with very little fidelity to the candidate in whose interest they pretend to be engaged, and if they have nothing more to receive, they immediately forget the presents they have already taken, and, without the least hesitation, espouse the cause of a more wealthy competitor.

'This pretended liberty, therefore, from which the Poles would be thought to derive so much glory, is no more than a slave to avidity. They sacrifice their privileges to repasts, or the purses of the candidates. They have been seen to crouch under the inglorious yoke of foreigners without making any effort to support the king they had elected, and they have abandoned their country as a prey to the Germans, whom they constantly hated, and likewise to the Russians, who were always a contemptible and conquered enemy in the reigns of Stephen Battori and John Sobieski.

'When any candidate has gained the suffrages of all the palatinates he is declared to be the elected king by the Archbishop of Gnesna, and is accordingly proclaimed as such by the marshals of the Crown and Grand Duchy, in conjunction with the nuncios. An oath is then exacted from the new monarch in favour of the *Pacta Conventa*, and when he has sworn to conform to the regulations of the diet with reference to the exorbitances, and to observe all the other laws of the kingdom, they proceed to the ceremonials of his coronation.

' Popery is the established religion, and the prince himself is obliged to profess it. The zeal of the Poles has always prevailed with relation to this article, and all the efforts made by the advocates for the Augsburg Confession for the introduction of a Protestant prince on the throne of Poland have hitherto proved ineffectual. There are, however, great numbers of Christians of the Greek Church, as well as Lutherans, Anabaptists, and Socinians, in this kingdom, and there are even Pagans to be found in the recesses of Lithuana. The people in general are more superstitious than devout, and they receive whatever is transmitted to them from Rome with a blind submission, and without bestowing the least examination on the particulars.

' The monks improve this stupid credulity to their own advantage; they frequently interfere in affairs of State, and enrich themselves by those means. The Jesuits of Leopold in Russia have a cope entirely covered with gold and precious stones in their treasury, but it is likewise so exceeding weighty that the priest is incapable of using it at mass. This ornament alone is valued at 50,000 crowns.

' With respect to the manners and disposition of this nation, the Poles, though they are naturally haughty and imperious, are yet sufficiently qualified to return any polite treatment they receive, and if a stranger only tenders them the first civilities, and is sedulous to cultivate their friendship, their behaviour will always correspond with his advances, and they will be industrious to render him all the good offices imaginable.

' Magnificence is the foible of the nobility, and they sacrifice all things to scenes of luxury. As they seldom behold any person superior to themselves in their own country, and as they treat their inferiors with an air of absolute authority, they live in all the splendour of princes, while fortune proves favourable to their inclinations. Prodigality and debauch are considered as virtues by a martial nobility, who are frequently precipitated into extremes by independence and impunity. Arms are their only occupation, they discover but little curiosity to cultivate the polite arts, and commerce is transacted among them by none but strangers. In a word, the Poles are only solicitous to distinguish themselves in war, to defend their frontiers, and to be vigilant over the conduct of their kings, since they think it degrading to indulge themselves in any other occupations. As an air of sincerity is diffused through all their conduct their friendships are constant, but they are easily rendered the dupes of their enemies. As they are very disinterested in their desires they seldom amass great riches, and frequently dissipate their patrimony. If they are at any time reduced to a melancholy state of indigence they borrow without any intention of making a proper restitution, and they think themselves privileged to dispose of the property of others with the same prodigality in which they waste their own. They appear serene and undisturbed amidst the greatest calamities, and behold the miseries of their friends and countrymen, and nearest relations, with an aspect of indifference equal to the insensibility with which they support their own misfortunes. They are naturally couragous and intrepid, and habituate themselves to all sorts of fatigue, and would indeed be invincible if they paid a due respect to their chiefs. This is the general character of the Poles, whose history I have undertaken to write.'

The history by the Abbé des Fontaines, brings us down to 1733, and tells of endless revolutions ; later history tells of partitions of the kingdom.

In an article in the *Edinburgh Encyclopædia* it is stated :—
'The Crown of Poland, with the exception of five centuries previous to the year 1370, was purely elective. Its sovereigns, whose authority before the era just mentioned was unlimited and absolute, were originally termed

duces, dukes, or generals, in reference to the almost invariable practice of their conducting the armies of the state to the field in person. In the fourteenth century the nobles availed themselves of the weakness of a female reign to diminish the power of their sovereign, and to extend that of their own order. They enacted that no taxes should be levied, that no new laws should be passed,—in short, that no measure of any importance should, as formerly, be effected by the king, but by representatives chosen from among themselves. Hence the origin of the diets of Poland, of which there were two kinds, Ordinary and Extraordinary, the former statedly assembled once in two years, while the latter was summoned by the king only on great emergencies. The diets consisted of the king, the senators, and deputies from provinces and towns, amounting altogether to about four hundred members. These assemblies could sit only for a limited time, and any individal, however humble, had the power of calling for a division of the meeting on any question, and one dissentient voice had the effect of rendering the whole deliberations ineffectual. This latter right, which was termed *liberum veto*, and which was repeatedly exercised, was the cause of the greatest calamities, and often of much bloodshed. Without the unanimous consent of the diet the king could determine no question of importance, could not declare war, make peace, raise levies, employ auxiliaries, or admit foreign troops into his dominions, with other restrictions, which almost extinguished the regal authority. Nor did the nobility stop here. Having thus undermined the power of the king there was but one step more to gain to them-selves the uncontrolled government of the nation, namely, to render the throne elective. This was accordingly accomplished, and the king of Poland enjoyed now the title, but little of the power or dignity of a free sovereign. Liberty, so much boasted of by the Poles, seems from this period to have been confined to the nobles alone. They arrogated an unlimited sway over their respective terri-

G

tories; some of them were hereditary sovereigns of cities
and villages, with which the king had no concern; they
exercised a power of life and death over their tenants
and vassals; they were exempted from taxes; and could
not be arrested and imprisoned but for a few crimes of the
basest kind. But the most dangerous of all their rights,
and one which made their situation analogous to that of
the German princes, was the power of constructing fortresses for their private defence, and of maintaining a military force, which, in imitation of real dignity, they caused
to keep guard round their palaces. The election of the
king was vested in them alone; and none but they, and
the citizens of a few particular towns, possessed the privilege of purchasing or inheriting property in land. The
nobility, amounting to about 500,000 individuals, Malte-
Brun emphatically terms the sovereign *body* of Poland.

'The senate, however, which owed its origin (in the
eleventh century) to Boseslaus I. formed an intermediate
authority between the king and the nobles. This body,
composed of the ministers of state, of the representatives
of the clergy, of palatines, and castellans, consisted of 149
members, until 1767, when four new members were added
to the number, as representatives of the province of
Lithuania. The senators, except the representatives of
the clergy, were nominated by the king, but continued in
office for life, and after their appointment were totally
independent of royal authority, to which, indeed, they were
regarded as a valuable counterpoise. The duty of the
senate was to preside over the laws, to be the guardians
of liberty, and the protectors of justice and equity, and,
conjunctly with the king, to ratify laws made by the
nobility. A diet could not, as previously hinted, be
constituted without the junction of the senate to the
national representatives; a portion of the senators,
indeed, acted as a committee for facilitating and conducting the public business of that assembly. The president of the senate was the archbishop of Gnesna, who,
during an interregnum, discharged the functions of king,

and enjoyed all the royal prerogatives but that of dispensing justice. His duty also was to summon the extraordinary diet when the throne became vacant, and to preside at that assembly.

'The diet which thus assembled on the death of a king of Poland to elect a sovereign to occupy the vacant throne was not unfrequently characterised by the most sanguinary proceedings. This assembly, which consisted of the senate, of the representatives of districts, of the clergy, and the nobles—the latter a most numerous body—met on horseback in a plain adjoining the village of Wohla in the neighbourhood of Warsaw. Though the electors were prohibited from appearing at the meeting attended by any body guard, they yet uniformly came armed with pistols and sabres, prepared to perpetrate the greatest excesses. Every member of the diet, as previously mentioned, was entitled to call for a division of the assembly on any question, or to put an end to the deliberations, or even the existence of the assembly, merely by protesting against its proceedings. This singular and absurd privilege, which was frequently exercised in those meetings of unenlightened and violent men, was productive of the most fatal consequences. It often led the stronger party to attack, on the spot, their antagonists, sword in hand; and it not unfrequently formed the origin of civil wars, by which the resources and the stability of the nation were undermined, patriotism extinguished, and the progress of liberal knowledge retarded. Before the successful candidate was proclaimed king, he had to sign the *Pacta conventa*, or the conditions on which he obtained the crown, which, on his knees, he had to swear never to violate.

'Such was the ancient constitution of Poland,—monarchy blended with aristocracy, in which, for several centuries previously to its dissolution, the latter prevailed. The Poles, indeed, denominated their government a *republic*, because the king, so extremely limited in his prerogative, resembled more the chief of a commonwealth than the sovereign of a monarchy. But it wanted one of

the necessary characteristics of a republic: The people
were kept in a state of slavery and vassalage, and enjoyed
not even the semblance of any civil privilege: the whole
power was engrossed by the nobles; and thus the Polish
constitution possessed not that community of interests,
that general diffusion of political privileges which are the
very life and stamina of a republic, as well as of a mixed
monarchy, and with which, in spite of much internal mis-
rule, and of the agressions of foreign enemies, Poland
might have flourished to this day.

‘The administration of justice, and the execution of the
law in Poland, was characterised by the grossest abuses.
The judges, nominated by the king, were chosen without
the most remote regard to their talents or integrity; the
decision of the courts of law were openly and unblushingly
sold to the highest bidder; and no cause, whatever its
merits, could be successful, unless supported by the all-
prevailing power of money. Nor was this corruption, for
which no redress could be obtained, confined to cases in
which the litigants were wealthy, or in which, as resulting
from some base and unprincipled transaction, the most
ample and liberal payment ought to have been demanded.
Actions for which a man deserved the thanks of the state,
were, when brought before a legal tribunal, the source of
much unjust expense to the person performing them.
“If a man apprehended a murderer,” says a writer quoted
by Malte-Brun, “and brought him before the proper
officer, he was charged ten ducats for his trouble, which,
if he were unable or unwilling to pay, the murderer was
immediately set at liberty.” Had he submitted to this
payment, the sums that would, on some pretence or other,
have been exacted of him, ere the offender was brought to
justice, no man unacquainted with the history of Poland
could conjecture. “It has cost,” says the same writer, “a
merchant of Warsaw 14,000 ducats for apprehending two
thieves.” Nor was the expense of a plea more to be
execrated than the duration of it. No litigation, even the
simplest—one, for example, between a debtor and credi-

tor—could be brought to a termination in less than four
years. In so short a time, however, was almost no case
decided. Vautrin mentions that he knew cases which had
been pending for sixty years, and which, so far as he could
foresee, might continue undetermined "till the last
generation." The nobles, disgusted with this fatiguing
tediousness, not unfrequently withdrew their pleas from
the decision of the courts of law, and settled them by
force of arms,—of which some instances occurred even so
recently as the middle of the last century.'

Such seems to have been the history of Poland. In
the fabulous period of Polish history the land appears to
have been colonised or conquered, and ruled by the
followers and descendants of a Scandinavian or a Caucasian
hero called Lekh, led thither by a white eagle, which
became the national symbol, and from whom the old
English name for the people—Polack, and the Russian
name Polekh, or Polach, was given to them.

The nobles were always peers, and equals in political
rank, and by some it is alleged that the whole population
were such for a long time, but the yeomen became serfs,
if they were not so originally. The monarchy was limited
in power, and became elective. At first successive concessions were made by each one ascending the throne, and
ultimately absolute freedom of election was proclaimed, an
annuity alone been required to render it valid, and a charter
was prepared, by which all privileges granted by previous
sovereigns were renewed and confirmed. It was established
that the king was to be chosen by the whole body of the
nobility and freeholders of the nation, and that in case of
the king infringing the laws and privileges, his subjects
should be absolved from their oaths of allegiance.

By union with Lithuania, and consequent on the various
changes, civil and ecclesiastical, which followed, curtailing
both civil and religious liberty, and ultimately in 1772
the first partition of the kingdom occurred.

There are in the list of the kings of Poland names which
have been held in high honour by the lovers of freedom.

The partition may be traced to measures adopted by the dominant party to curtail the religious liberty of dissenters from the Church of the State and the Church of Rome. 'The reformed religion, though early introduced into Poland, was not for two centuries very generally adopted. The Protestants, called *Dissidents* (a term which also comprised those of the Greek church), were tolerated, though they were obliged to labour under many civil disabilities. During the interregnum that preceded the election of Poniatowsky, a decree had been made by the diet, by which the dissidents were, in a great measure, forbidden the free exercise of their worship, and totally excluded from all civil and political privileges.

. . . .

'The dissidents could not submit without a struggle to the deprivation of their most valuable privileges. They combined unanimously to endeavour to accomplish the repeal of this decree, and, for this purpose, applied for advice and assistance to some of the most eminent powers of Europe. And accordingly Russia, Prussia, Great Britain, and Denmark, made remonstrances to the Government of Poland on this subject. These remonstrances, however, were without effect; for the decree was confirmed by the coronation diet held after the king's election. The dissidents in the meantime presented to the government petitions and memorials, and the decision of the question was at last referred by the diet to the bishops and senators. And upon a report from them, the diet made some concessions, which, however, were far from satisfying the dissidents, who thought it absurd that the redress of their grievances should be entrusted to those very persons who were the *authors* of them. The dissidents, whose cause was now openly espoused by Russia, Prussia, and Austria, were not to be flattered by the concessions of these persecutors, nor overawed by their power. They formed confederacies for their defence in every province, and were determined to resist unto blood in support of their rights and privileges. Nor were the Popish clergy and their adherents slow in making preparations. *The Confederation of the Barr*, the hope and bulwark of their party, took up arms. The cries of liberty and religion became every where the signal of a war, the true object of which, with the Catholics, was, not only to disperse or destroy their opponents, but to dethrone Stanislaus, whom they regarded as friendly to the dissidents, and to rescue Poland from the influence of Russia. The *confederates*, as the Catholics were now termed, feebly supported by Saxony and France, were vanquished in almost every battle ; and the dissidents would have been secured in the open and unshackled profession of their faith, had the sovereigns to whom, in no mean degree, they owed their success, been actuated by any regard to their cause, or had not trampled under foot every principle, which the law of nations,--which the law of nature, should have taught them to cherish and reverence.

' These sovereigns, however, instead of being animated in the cause of civil and religious liberty, were, under the false pretences, labouring solely to extend the boundaries of their respective dominions, and to promote the aggrandisement of their power. Nothing less than the dismemberment of Poland, and the partition of it among themselves, was their object in the assistance they afforded the dissidents,—an object which could only be attained, or at least more easily attained, by fomenting internal divisions, and thus undermining the resources and unanimity of the kingdom. This plan, it is thought, was first contemplated by Prussia ; but Russia and Austria readily enough embraced it, though all these kingdoms at different periods owed much of their glory, and even their very existence, to the country which they thus resolved to destroy. A great proportion of Poland was thus seized upon by these kingdoms, and a treaty to this effect was signed by their plenipotentiaries at St. Petersburg in February 1772. The partitioning powers having forced the Poles to call a meeting of the diet, threatened, if the treaty of dismemberment was not unanimously sanctioned, that the whole

kingdom should immediately be laid under military execution, and be treated as a conquered state. The glory of
Poland was past; and though some of the nobles, rather
than be the instruments of bringing their country to
ruin, chose to spend their days in exile and poverty, the
measure was at length agreed to; and Stanislaus himself,
threatened with deposition and imprisonment, was prevailed upon to sanction it. Europe, though astonished at
what was taking place in Poland, remained inactive. The
courts of London, Paris, Stockholm, and Copenhagen,
indeed, sent remonstrances against this usurpation; but
remonstrances without a military force will, as in the case
before us, be ofttimes unavailing.

> "Oh bloodiest picture in the book of time,
> Sarmatia fell, unwept, without a crime!"

A large portion of the eastern provinces were seized by
Russia; Austria appropriated a fertile tract on the south-
west; while Prussia acquired a commercial district in the
north-west, including the lower part of the Vistula.
Poland was thus robbed of 70,000 square miles, or about
a fourth of her whole territory.'

Wars followed. Russia and Prussia triumphed. Warsaw
still remained unconquered. At length Warsaw also fell
before an invading host.

'Poland being thus overthrown, the two ursurping
powers were about to form a partition of it betwixt them,
when Austria unexpectedly stept forward, and declared
that she could not permit the entire destruction of Poland,
unless she were allowed to share in the division. The
consequences of a refusal they were not willing to encounter; and Austria had thus her ambitious views realised,
without having incurred the smallest danger or expense.
Stanislaus, who had all this while remained in his capital,
was at length removed to Grodno a second time, where he
was compelled to resign his crown, and was thence carried
to St. Petersburg, where he resided as a state prisoner in

solitude and exile till his death, which took place in February 1798.

'The result of this partition was as follows :—

	Square Miles.	Population.
To Austria, ...	64,000	4,800,000
To Prussia, . : . :	52,000	3,500,000
To Russia, . . .	168,000	6,700,000
	284,000	15,000,000

'Of this territory, the partitioning powers appropriated to themselves those districts that lay most convenient to their dominions, the acquisitions of Russia being larger than those of the other two taken collectively.'

On this occasion, as was remarked at the time, the largest share went to Russia, the most populous to Austria, and the most commercial to Prussia. The portion which fell to Russia contained numbers of inhabitants who were already connected with that country by religious ties ; and I may take occasion to state that while I have heard very severe things said by Polish noblemen against Russia, I have been told by a Pole that there exists in Poland a still more bitter feeling against Prussia.

The conquests of Napoleon made great changes in Poland ; and at the great settlement of 1815 the Emperor Alexander I. proposed to form the whole of ancient Poland into a constitutional state under the Russian crown ; but it was ultimately arranged that Galicia should be given back to Austria, Posena to Prussia, and that the rest of the Napoleonic Duchy should be formed into a constitutional state, with the Emperor of Russia as king — provinces acquired by Catherine II. at the portion of 1772 remaining incorporated with the Russian empire ; and it was governed accordingly. It had its diet, its national administration, and its national army. In 1830 an insurrection broke out. When this was suppressed all that was changed, and measures were taken to Russianise

the country if possible. A second insurrection broke out
in 1863, and this also was suppressed.

The population is reported as about five millions of
souls. I presume of inhabitants, though in Russia that
term is applied to males alone. According to one estimate reported to me, about two-fifths of the inhabitants
are Jews, and something less than that proportion are
Poles. According to a Russian estimate, the Russians
are under 10,000 males, and the Jews number upwards
of 600,000 males.

The serfdom of the peasantry was comparatively
restricted, but it could be made very oppressive. In the
insurrection of 1830 the leaders sought to secure their co-
operation by promising them emancipation and a freehold
possession of lands in their occupation; and a similar
arrangement was adopted by the Russian Government.

I have not the information necessary to enable me
to form an opinion in regard to the propriety, expediency,
or the justice of subsequent proceedings, which brought
ruin and exile upon many noble-minded natives of Poland,
males and females, nobles and peasants, alike. In the
minds of many, Poles and Siberia are closely associated.
I have not the means of either verifying or disproving
any of the allegations which are current in regard to the
number of Poles who have been expatriated by exile
thither. But to this I can testify: for about seven years—
from 1833 to 1840 inclusive—every male going to Siberia
was supplied with a copy of the New Testament scriptures.
All of these passed through my hands; and in no one year
did the number of New Testaments in Polish appear to
be disproportionate to the numbers required in other
languages spoken in the provinces — Finnish, Lettish,
Estonian, Swedish, German, and French.

Of the Polish exiles on their journey to Siberia,
Michie, in his volume entitled *The Siberian Overland
Route from Pekin to St. Petersburg*, mentions that the
number of Polish prisoners met with on the road

at some places threatened seriously to impede his
journey. Between Kazan and Perm he encountered
companies of them on their way, and this at almost
every station. This was in 1863 or 1864, and he writes :—
' The resources of the posting establishments were
severely taxed to provide horses for so many travellers at
once, and we had frequently to wait till the Poles were
gone, and then take the tired horses they had brought
from the last station. The Poles travelled in the same
manner as we did, in large sledges, containing three or
four people, sometimes more. Those who could not be
accommodated with sledges had carts or *telégas*, which
were more or less crowded. None of them travelled a-foot.
All were well clothed in furs. On the whole, I was surprised to find such a number of people travelling with so
much comfort. The prisoners were invariably treated
with kindness and consideration by the officer in charge
and by the *gensdarmes*. They are under close surveillance,
but I did not see any of the prisoners in irons, though I
was informed that some of them were so. . . . They
ate well, and talked loudly ; the din of their voices at a
post-station was intolerable. Many joked and laughed a
great deal, by way of keeping up their spirits, I suppose ;
but no indication whatever was given that they were exiles
undergoing the process of banishment. If one might
judge from appearances, I should say they rather liked it.'
Of the passage of the Volga, he writes :—' The ferry-
boats were engaged the whole day in conveying Polish
exiles across the river bound for Siberia. It is a sad sight
to see so many people in captivity, and still more so to see
a number of women accompanying the exiles. It is quite
common for the wives, daughters, and mothers of the poli-
tician convicts to follow their relatives into Siberia. This
is not discouraged by the Russian Government ; on the
contrary, every facility is granted to enable their families
to emigrate, and they have always the means of travelling
in company. The object of the Government is to colonise
Siberia, so that the more people who go there the better.

Besides, the residence of families in exile offers some
guarantee against any attempt at a return to their native
country.' Describing then two old ladies, well dressed
in black silk and warm fur cloaks, he says:—' They were
treated with great kindness by the soldiers, who lifted
them carefully out of the boat, carried them to their
sledges which were in waiting, and put them in as tenderly
as if they had been their own mothers. After carefully
wrapping them up with their furs a Cossack got in beside
each of the ladies, and they drove off to Kazan. A girl
who was with them was equally well attended to by the
officer in command of the party, who seemed to consider
the Polish maiden to be his especial charge.'

Some additional information relative to the opinions
entertained by Polish exiles in Siberia and others
in regard to advantages which they enjoy in Siberia
is supplied by Mr Michie (pp. 337-342). I believe
it to be a fair representation of the case; and it is in
accordance with everything stated by Dr Lansdell in
his volume entitled *Through Siberia*, published in
1882, in which he gives details of arrangements of almost
every prison in Siberia, to visit which, with a view to
supplying the prisoners with copies of the New Testament
was the design of his journey. A review of what he says
I embodied in a companion volume to this, entitled
*Forestry in the Ural Mountains in Eastern Russia.**

* *Forestry in the Mining Districts of the Ural Mountains in Eastern Russia.* In
which are given details of a journey from St Petersburg thither, and of forest exploitation in the government of Ufa; an account of the Ural Mountains and the population
of metallurgy works creating a demand for forest products; an account of the forests
of the district; of the exploitation of these; and of abuses connected with this, with
a parting glimpse at the life of the people; and the conquest of Siberia by Russia.

PART II.

LITHUANIA.

————:o:————

CHAPTER I.

LITHUANIA AND ITS PEOPLE.

ACCORDING to a Gazeteer published in the last century (1798), Lithuania is 'a large country between Poland and Russia. It is about 300 miles long, and 250 in breadth. It is watered by several rivers, the principal of which are the Dneiper, the Dwina, the Nieman, the Pripeez, and the Bog. It is a flat country like Poland, and the lands are well adapted for tillage. The soil is not only fertile in corn, but it produces heavy wood, pitch, and vast quantities of wool. They have also excellent little horses which they never shoe, because their hoofs are very hard. There are vast forests, in which are bears, wolves, elks, wild oxen, lynxes, beavers, gluttons, wild cats, &c., and eagles and vultures are very common. In these forests large pieces of yellow amber are dug up frequently. The country abounds with Jews, who, though numerous in every other part of Poland, seem to have fixed their head quarters in this duchy, and this is perhaps the only country in Europe where Jews cultivate the ground. The peasants are in a state of the most abject vassalage. The religion was formerly Romish, but now there are Lutherans, Calvinists, Socinians, Greeks, and even Turks, as well as Jews.'

As it was then, so is it still. In a later Gazeteer published by Fullarton, it is stated :—' Lithuania, called in German Littauen, is a very ancient division of Europe, lying between Courland, Russia, Poland, and Prussia. In the eleventh century the country was tributary to Russia; in the thirteenth it became a grand duchy under Ringold. The Grand Duke Jagellan, having married the Polish princess Hedwig, united his duchy to the crown of Poland about the year 1386. At the first partition of Poland in 1773, a considerable portion of Lithuania was assigned to Russia, and the governments of Mohilev and Polozk, or Vitebsk, were formed out of the newly acquired territory; while the remainder of Lithuania, forming the territories of Vilna, Troki, Polozk, Novogrodek, Brzesc, and Minsk, remained attached to the Polish monarchy. By the partition of 1793 and 1795, Russia further acquired those portions of Lithuania which now form her governments of Vilna, Grodno, and Minsk, while Prussia acquired that portion which constituted her regency of Gumbinnan.' It is the Russian portion, exclusive of this, upon which I now report.

Lithuania, like Poland, has played an important part in the development of Russia, in so far at least as modification of political government is concerned. In Russia at the present time, we have, in combination with absolute monarchal power in all that relates to national matters, what looks like absolute democratic power in all that relates to communal matters, and in a former day there was in Russia in one district a democratic republic side by side with an absolute monarchy. According to one legend the real founders of the Russian monarchy were Normans who found their way to the district in which is situated the city of Novgorod, either coming thither by invitation or in the prosecution of their personal pursuits.

Of what followed Mr Mackenzie Wallace writes :—
' For six centuries after the so-called invitation of Rurik the city on the Volkhof had a strange chequered history. Rapidly it conquered the neighbouring Finnish tribes, and grew into a powerful independent state, with a territory extending to the Gulf of Finland, and northwards to the White Sea. At the same time its com-

mercial importance increased, and it became an outpost
of the Hanseatic League.   In this work the descendants
of Rurik played an important part, but they were always
kept in strict subordination to the popular will.   Political
freedom kept pace with commercial prosperity.   What
means Rurik employed for establishing and preserving
order we know not, but we know that his successors in
Novgorod possessed merely such authority as was freely
granted them by the people.   The supreme power
resided, not in the prince, but in the assembly of the
citizens called together in the market-place by the sound
of the great bell.   This assembly made laws for the
prince as well as for the people, entered into alliances
with foreign powers, declared war and concluded peace,
imposed taxes, raised troops, and not only elected the
magistrates, but also judged and deposed them when it
thought fit.   The prince was little more than the hired
commander of the troops and the president of the judicial
administration.   When entering on his functions he had
to take a solemn oath that he would faithfully observe
the ancient laws and usages, and if he failed to fulfil his
promise he was sure to be summarily deposed and
expelled.   The people had an old rhymed proverb, " *Koli
khud knayaz, tak v grayaz !*" ("If the prince is bad, into
the mud with him !"), and they habitually acted according
to it.   So unpleasant, indeed, was the task of ruling those
sturdy, stiff-necked burghers, that some princes refused to
undertake it, and others, having tried it for a time,
voluntarily laid down their authority and departed.   But
these frequent depositions and abdications—as many as
thirty took place in the course of a single century—did
not permanently disturb the existing order of things.
The descendants of Rurik were numerous, and there were
always plenty of candidates for the vacant place.   The
municipal republic continued to grow in strength and in
riches, and during the thirteenth and fourteenth century
it proudly styled itself " Lord Novgorod the Great"
(*Gospodin Veliki Novgorod*).

'" Then came a change, as all things human change."
To the East arose the principality of Moscow—not an
old, rich municipal republic, but a young, vigorous State,
ruled by a line of crafty, energetic, ambitious, and un-
scrupulous princes, who were freeing the country from
the Tartar yoke and gradually annexing by fair means
and foul the neighbouring principalities to their own
dominions.   At the same time, and in a similar manner,
the Lithuanian princes to the Westward united various
small principalities, and formed a powerful independent
State.   Thus Novgorod found itself between two powerful
aggressive neighbours.   Under a strong government it
might have held its own against these rivals, and successfully maintained its independence, but its strength was
already undermined by internal dissensions.   Political
liberty had led to anarchy.   Again and again on that
great open space where the national monument now
stands, and in the market-place on the other side of the
river, scenes of disorder and bloodshed took place, and
more than once on the bridge battles were fought by con-
tending factions.   Sometimes it was a contest between
rival families, and sometimes a struggle between the
municipal aristocracy, who sought to monopolise the
political power, and the common people, who wished to
have a large share in the administration.   A State thus
divided against itself could not long resist the aggressive
tendencies of powerful neighbours.   Artful diplomacy
could but postpone the evil day, and it required no great
political foresight to predict that sooner or later Novgorod
must become Lithuanian or Muscovite.   The great families
inclined to Lithuania, but the popular party and the
clergy looked to Moscow for assistance, and the Grand
Princes of Muscovy ultimately gained the prize.'

There are indications in early legends and history that
the most absolute liberty was enjoyed, or let me rather
say, possessed by the Russian peasantry.   Not only was
every man's house his castle, but every man was king in

his own family, and was the subject of no one. At a later period the so-called noblemen were nominally and to some extent under the authority of the prince, but it was a subjection which left the man free to do as he liked. By degrees the prince gained power, and the nobles lost it. And in reference to the changes going on in the times just alluded to, Mackenzie Wallace writes:—' When the Grand Princes of Moscow brought the other principalities under their power, and formed them into the Tsardom of Muscovy, the nobles descended another step in the political scale. So long as there were many principalities they could quit the service of a prince, as soon as he gave them reason to be discontented, knowing that they would be well received by one of his rivals ; but now they had no longer any choice. The only rival of Moscow was Lithuania, and precautions were taken to prevent the discontented from crossing the Lithuanian frontier. The nobles were no longer voluntary adherents of a prince, but had become subjects of a Tsar ; and the Tsars were not as the old princes had been. By a violent legal fiction they conceived themselves to be the successors of the Byzantine Emperors, and created a new court ceremonial, borrowed partly from Constantinople and partly from the Tartar Horde. They no longer associated familiarly with the Boyars, and no longer asked their advice, but treated them rather as menials. When the nobles entered their august master's presence they prostrated themselves in Oriental fashion—occasionally as many as thirty times—and when they incurred his displeasure they were summarily flogged or executed, according to the Tsar's good pleasure. In succeeding to the power of the Khans, the Tsars had adopted, we see, a good deal of the Tartar system of government.'

To this has to be added the impoverishment of the nobles through extravagant expenditure, and the privations of the poor through lack of remunerative employment, and, it may be, want of energy and past wrongs done to them by petty local officials. The last mentioned evil

H

is one not limited to this province, but an illustration taken from it may be more satisfactory than mere illusions to the general forms which it assumes throughout the empire.

By one of my personal friends I was informed that on the occasion of wide spread famine in this region some years ago, provision was made by the Government for the supply of the destitute with corn. He happened to be present when the Government official, appointed to enquire into the state of the poor in the locality where he was, came to the house of a poor man whom he knew to be in want. To his surprise, in answer to the enquiry of the official whether he needed help, he said, ' No.' The official seemed equally surprised, and varied the form of the question, but still got the same reply. He began to reason with the man in regard to the folly of not availing himself of the provision which the Government had made for supplying corn to them free of all charge; but still the man maintained that he did not need it. And the apparent philanthrophy of the official was completely baffled by the apparent independent spirit of the peasant. When the official had gone my friend astonished asked of him, ' How is this? You and your family are in danger of starvation, and yet you say you are not in want. I cannot understand it.' ' Yes,' said the peasant, ' we are in want. But of whatever grant may be made by the Government a great deal will be retained by those through whose hands it first passes ; the same will be done again and again by all those through whose hands it must pass in succession before it could reach us ; and reach us it never will. So if we must starve any way I would rather starve just as I am than give to these fellows an opportunity of making money under the pretext of saving us from starvation.'

But among the courtiers there was to be found, in the times spoken of by Mackenzie Wallace, or shortly thereafter, ' Lithuanian nobles, who found it more profitable to serve the Tsar than their own sovereign.'

Of the treatment to which men of their order were subject twenty years ago, a picture is given by the Rev. Fortescue L. M. Anderson in a little volume entitled *Seven Months Residence in Russian Poland in 1863*.

In 1863 the Rev. Fortescue L. M. Anderson accepted an invitation from Count Alexander von Bisping-Galen to accompany him to his estate of Wereiki in Lithuania, he having made his acquaintance at the University of Bonn, where his father held the appointment of English chaplain. The Count was in his political views what was known as a Conservative Pole, that is one who was opposed to insurrection ; and as indicative of the desire of the Count to avoid even the appearance of sympathy with the disaffected it is mentioned by Mr Anderson, in giving a narrative of their journey, in prosecuting which they had reached Konigsberg, 'he (the Count) had bought some time before a pair of guns and pistols of superior workmanship, which he naturally wished to carry home with him. But to have tried to introduce these with his own luggage at the Russian frontier would have been to expose them to certain seizure, and himself probably to heavy penalties. To have sent them (as he might have done) by boat up the river, along with a reaping machine which he had bought in England, and which was now awaiting his arrival in Konigsberg, would have been simply an evasion of the law, and have subjected him to the charge of secretly supplying arms for the insurgents. He resolved, therefore, at once to abstain from any and every attempt, directly or indirectly, to make use of the guns and pistols which he prized so highly; and begged me to ask my father, in the letter I was then writing, to take charge of them. They were accordingly sent off to Bonn, before our departure from Konigsberg; and my father has them still in his possession, with the cases unopened.'

They found everywhere an abundance of military, but the first offensive conduct which they saw was at Grodno. 'From Wilna our next journey was to Grodno; the

country is for the most part flat, and only diversified here
and there by small lakes, and patches of fir and birch
plantations.   The intermediate stations were all crowded
with soldiers ; but no authentic particulars of information
could be learned anywhere respecting the insurrection
which had thus drawn them out from their various
winter quarters.   We had learned much more about its
progress, whilst we were yet at a distance, from English
and French and German newspapers, than we seemed
likely to learn in the country disturbed by it.

'A run of about four hours by railway brought us to
Grodno, where we were received by a cousin of the
Count ; and, having dressed hastily at the hotel, went to
dine with Prince and Princess Lubecki, who received us
with the utmost kindness.   I was agreeably surprised
to find that the Princess, who is an aunt of the Count,
spoke English ; and this circumstance, joined with the
hearty friendliness of all the party, tended not a little to
secure for me a most agreeable evening in this land of
strangers.

'Our breakfast next morning presented the novelty of
tea served up in tumblers, accompanied with cigarettes,
and, from that time until midday, there was an uninter-
rupted succession of visitors coming in to see the Count.
We went out for luncheon to a neighbouring restaurant,
and met a number of Russian officers who had been
playing billiards there.   The demeanour of the Russian
officers, especially those of the Imperial Guard, is generally marked by great courtesy ; but, upon the present
occasion, I must confess, they exhibited it only in scant
measure.   In fact, one of them did not scruple to stretch
his hand rudely over the table and help himself to a
portion of bread which had been placed for our party.
Whether he did it intentionally, or from inadvertence, I
cannot say ; but a significant smile which passed, at the
same moment, over the face of one of his brother officers
who observed him, forced me to believe that they were
trying how far they might venture to show their contempt

for those whom they regarded as merely Polish intruders.
But their plan to pick a quarrel with us—if such had
been their purpose—was frustrated by our taking not the
slightest notice of their conduct.

' On Friday morning, the 6th of March, we started by
post-waggon for Wereiki, one of the Count's country
houses, south of the Niemen, about thirty English miles
from Grodno. The two waggons, which drew up to the
door an hour after the appointed time, resembled Scotch
hay-carts. One of them was drawn by four rough-looking
horses abreast, and the other by three. The rapid
pace at which they galloped along the wretched roads
was perfectly surprising, and the violent jolting which the
passengers thereby suffered, for the waggons had no
springs, may be better imagined than described. Our
progress at first was slow enough ; the rough pavement
of the streets making it necessary to traverse them
nearly at a walking pace, and the obstruction at the
bridge over the Niemen being not easy to overcome.
There are two bridges across this part of the Niemen ;
the one, a tubular bridge, for the railway to Warsaw ;
the other, a flying bridge, as it is called, but of a
very different construction from those bearing the
same name, with which the traveller on the Rhine is
familiar at Bonn, Konigswinter, and Neuwied. The
movement of the Rhine bridges, as the reader probably
knows, is carried on by the action of the stream, and
controlled by a chain, which is attached to the stern of
the two barges which support the platform of the
bridge, and thence passing through a high wooden frame-
work, is drawn over a series of boats stationed at intervals up the stream, and is fastened over the stern of the
hindmost boat, to the bottom of the river. The Niemen
bridge is likewise set in motion by the stream ; but,
instead of being attached to boats, its course, from one
bank to the other, is controlled by a stout cable, which
passes upon rollers, through two upright posts fixed on
each side of the bridge, and is fastened at each end to a

windlass on either side of the river. The Niemen is not
half so broad at this point as the Rhine, or this mode of
making the bridge swing to and fro would not be practicable. The platform, which composes the floor of the
bridge, is laid over two strong barges, like those upon
the Rhine, and large enough to carry, at each trip, carts
and horses and foot passengers. Upon the present
occasion there could not have been less than a hundred
passengers and fourteen or fifteen carts and waggons,
besides our own. This motley assemblage, packed of
course very closely together, presented a scene of hopeless confusion. Not much inconvenience, indeed, was
suffered by our own party, for precedence was given to
the post-waggon in which we were carried. But those
who followed could only gain the bridge by a general
scramble; men, women, and children, soldiers and peasants, pushing and quarrelling, swearing, screaming—
each striving to reach it first. Two of the party
especially attracted my attention, a peasant boy, and a
Jew woman about fifty years old, each leading a horse
and a four-wheeled market cart down towards the bridge.
The woman was rather in advance at first, but the boy,
whilst she was wrangling with the manager of the bridge,
contrived to get his horse and cart in front of hers.
Whereupon the woman, as soon as she perceived it,
beat his horse savagely about the head, and then turned
with equal fury upon the boy himself, scratching his
face and kicking him. The boy was not slow to pay
her in her own coin, and a regular pounding-match
followed, to the amusement, apparently, of the lookers-
on. But, in the midst of the scuffle, the boy was still
mindful of his main purpose, and, watching his opportunity, succeeded in making good his stand upon the
bridge. The woman, too, followed, and would fain
have renewed the contest; but, finding this impracticable, contented herself with pouring out an incessant
torrent of abuse upon the object of her rage. The
so-called management of the bridge, as indeed is the

management of everything else in this country, is in the hands of Jews. But, if the preservation of order be part of the contract, it was certainly set at nought this day. There may have been officers present, whose duty it was to preserve order, but I could not see any. Indeed, it would have required a regiment of policemen to have withstood the pressure of the eager multitude.

'Once across the bridge, our pace was a continual gallop. Even a steep hill, which we had to ascend soon after leaving the river, and heavy roads, made still heavier by muddy heaps of half-melted snow, had no effect in slackening the speed of the poor horses. The driver was unceasingly employed in shouting to them, or in lashing their jaded sides with a rope whip. The first stage, about fifteen English miles, through a hilly country, was completed in less than an hour and a half; the second, about the same distance, was traversed in about the same time, over a somewhat better road, and the country was more flat. Our luncheon, at the end of the first stage, consisted of brown bread and cheese with caviar and claret. There was a long building by the road-side, which served as a stable for the post-horses, four of which are required always to be kept in readiness for any emergency. Boards were laid in a sloping form for the horses to lie upon, but covered with a very scanty littering of straw ; and little or no pains appeared to be taken to keep either stables or horses clean.

'We passed some good-looking farm-houses, in the second stage of our journey, and also some rich-looking fen land, which reminded me of the best parts of the border country of Norfolk and Lincolnshire.'

They made a round of visits to the relatives of the Count. All is described, with notices of the domestic habits of the families visited, with a naivete which shows that nothing political, much less anything treasonable, was in their thoughts.

While thus spending their time, and devoting much of their leisure to fishing and the chase, the insurrection which was being expected, broke out ; and Mr Anderson records some of the cases of what he considered oppression which came under his notice, when he went with his friend, Count Bisping, to spend the summer months with him on his family estates in Lithuania, little imagining that like treatment was to be given to themselves. But such was the case. He was about to return to Bonn, where he was expected before the end of September, when, most unexpectedly, he and his host together were thrown into prison. Of his arrest and imprisonment he tells :—
' On Friday, the fourth of September, we came to Grodno from Wereiki, intending to proceed the following week to Wiercieliszki, and attend the harvest home which was to be there celebrated. We remained in Grodno the whole of Saturday and Sunday; and, on Monday afternoon, set out for the farm. The weather was beautiful; the Jew post-master had sent us his best team of four horses to convey us; and the Count, his German servant, and I left the hotel, about three o'clock, in high spirits; little dreaming of the events that were soon to befall us.

' Upon reaching the town barrier, where the passports are always examined, I saw two sentries of the Imperial Guard standing in their dress uniform, in honour of some grand festival. The Count hurried upstairs as usual, with the papers, whilst I remained in the carriage. Hardly had he left me, when a *gensdarme* came up, and asked whether the carriage in which I was sitting belonged to Count Bisping. On my replying in the affirmative, he told me to go upstairs with the luggage, which consisted of two portmanteaus and a carpet-bag. I was somewhat startled at receiving such an order; but, without making any remark, at once obeyed it. Count Bisping was even more astonished at seeing me enter the passport office, and asked what was the matter? I referred him for information to the *gensdarme*, who forthwith showed, that, in his opinion, something very serious was the matter; and that it was his duty to call us strictly to account for it. In a loud, dictatorial tone, he ordered the servant to open the boxes, and demanded from us our papers. Both these orders were instantly complied with; but, in surrendering my papers I took the liberty of retaining my English passport as long as I could.

'After the boxes had been examined, and every article of their contents most minutely scrutinised, the officer turned to Count Bisping, and ordered him to undress. The Count immediately pulled off his coat and other garments. The contents also of his pockets were searched; his watch and purse seized, and all his money counted; even his boots were taken off and carefully examined by a soldier, to see if any piece of paper were concealed inside them, or between the soles. The German servant next underwent the same operation; whilst I quietly looked on. My turn then came. The *gensdarme* turned to me, and said, " Now, undress."

'" I decline doing so," I answered, speaking in French, " I am an Englishman, travelling with a passport given under the hand and seal of our Foreign Secretary; and I have also a Russian passport. I have not broken any of your laws; and, until I am informed of the cause of my arrest, I shall not submit to be treated like a felon."

'He stared at me with a look of blank amazement; apparently unable to believe his ears, that any one, connected with the despised Poles, should dare to disobey the orders of a Russian official. He paused, but only for a moment: then ran, and opened the window, and spoke some words of loud command, the purport of which, the Count told me, was to summon the *Chef militaire*, and the *Chef de police*, together with a troop of Cossacks, to our quarters. I could hardly help smiling at this alarming array of force, ordered out against one defenceless man; and was curious to see whether it were really intended by the Russian authorities to mark their proceedings by an act of such egregious folly.

'During the absence of the messenger, the Count said to me, " Why do you not yield to their demands ? "

'" I will do so," I replied, " as soon as I am informed what their demands are. But I will not submit to the indignity of being treated in this manner, without any reason assigned."

'The reader may probably think that I was wrong in making this resistance ; and, if more time for reflection had been afforded to me, I should probably have abstained from the attempt to interpose any delay. But, at the moment, I was filled with deep indignation, which absorbed every other thought.

'After waiting about half an hour, the *Chef militaire* arrived, and was soon followed by the *Chef de police* and the Cossacks. The latter were forthwith despatched at full gallop to the Count's farm, with orders to seize and bring away everything which they could find belonging to us. The *Chef militaire*, I must here mention, was an officer of the Imperial Guard, only acting temporarily in the office which he then filled. He was a thorough gentleman ; and I could read in his countenance the disgust which he felt at the arrogant offensive manner of the *Chef de police*. Having been informed of all the particulars of our arrest, and of my refusal to undress, the *Chef de police* turned round to me, and said, in broken English, " But you moost, my friend, you moost *déshabiller*." I repeated what I had before said, and told him, that, if the cause of my arrest were explained, I would obey any and every order which it was his duty to impose : but that, otherwise, I should do nothing. They had of course the power to do what they liked. It would be useless for me, I said, to resist them : and I could only appeal to the Governor for redress.

'" May the soldiers undress you ? " was the next question.

' "I have told you already that you may do what you like. I have no means of preventing you ; but I shall certainly not do anything to assist you in the infliction of an insult against which I protest. I have here my English passport, bearing the signature and seal of Lord Russell; and, if you look at it, you will see that it requires, *in the name of the Queen of England, all those whom it may concern, to allow me to pass freely without let or hindrance, and to afford me every assistance and protection of which I may stand in need.* Do you suppose that these words have no meaning? or that I shall quietly suffer you or any man to trample under foot the authority which they assert ? "

' "Let me see your passport," said he. I held it out to him. He seized it, crumpled it up, and thrust it into his pocket, adding in a contemptuous tone, " Well, now you have it no longer." I remonstrated as strongly as I could, against such cool audacious insolence ; but the only notice he took of me was to draw the passport out of his pocket again, straighten it, and place it among my other papers, which he duly sealed up.

' He then made a sign for two of the soldiers to draw near and strip me of my coat and waistcoat and boots, and examine carefully every part of them. Upon seeing them peep into my boots, I told the *Chef de police* that Englishmen were not in the habit of carrying money or letters in their boots. " Oh, no," he said, " I have been in England ; and know the ways of Englishmen well enough."

' " If that be the case, I think you must have learned that no man in England is treated as a criminal, until his crime be proved. Why have you not remembered the lesson ? "

' The work of undressing me being ended, the men proceeded to repack our portmanteaus, and left me sitting where I was, occasionally looking at me, and wondering apparently what was to happen next. As soon as everything was ready for removal, the *Chef de police* turned round, and said to me, "Come, Sir, dress."

' " No. I shall do nothing. You have brought me into
this condition, by what I regard as most unjust conduct.
I shall leave you to get me out of it as you can."

' Upon this the German servant was directed to do
what was required; and, had I not helped him from
time to time, I believe his trembling hands would never
have accomplished the task.

' " Why is the *pauvre Anglais* so cross ? " said the *Chef
de police*, in a taunting tone, to the Count. " He seems
quite angry."

" I am angry," I replied, " at the violation of justice,
which is committed under the name of law."

To like treatment he was subjected for many days.
Of this full details are given—pp. 142-214—but the
narrative is too long for citation. His deliverance was
the result of the circumstance of three English tourists
happening to pass through Grodno, who heard there that
one of their countrymen was imprisoned. They were the
Rev. W. G. Clark, Fellow and Tutor of Trinity College,
Cambridge, and public orator ; Mr W. Lloyd Birkbeck,
Fellow of Downing College, Cambridge ; and the third a
member of Balliol College, Oxford. Mr Clark wrote to
Lord Napier, British ambassador at St. Petersburg, and
Mr Birkbeck to Mr White, the Vice-Consul at Warsaw ;
and thus was secured his liberation. But his host was
exiled to the frontier of Siberia. Of his ultimate fate
Mr Anderson writes :—

' Little more remains to be told of what has since
befallen my dear friend, Count Bisping. The first intelligence which reached me respecting him was that he
had been set free. I was not at all surprised at this ; for
I knew that his arrest had produced the greatest excitement, and that there was a wide and deeply felt
sympathy among all classes for him ; and that every
effort was made in Grodno and throughout its neighbourhood to effect his release. A day or two before my
departure I heard that a long train of his tenantry and

neighbours, headed by a Russian priest, were coming to
the Governor of Grodno with a petition on his behalf.
The word of a Russian priest spoken in favour of a
Polish Roman Catholic proprietor was no ordinary incident; and I was sanguine in the hope of a successful
result. He was indeed released, but only for a short
time. He was still forbidden to leave Grodno; and, in
a few days afterwards, the hand of power was once more
laid npon him, and he was removed to St. Petersburg, on
his way to the distant province of Orenburg, which
adjoins the south-western range of the Ural Mountains.
He was told, indeed, that upon arriving at the place of
his exile, he should be allowed his personal liberty; that
his valet and man-cook should accompany him; that he
should have the command of his money for the supply of
what was needful; and that his estates should all be
preserved. Upon arriving at St. Petersburg he was again
placed in confinement for eight days, until the arrangements were completed for the subsequent disposal of himself and of those who were forwarded, at the same time,
to undergo the like or severer banishment.

'During his stay in St. Petersburg, his servant, who
had thus far continued to wait upon him, with the
intention of going on even to the end, went to the
Prussian Embassy, and their received such formidable
accounts of the risks to which (through ignorance of
the Russian language), he might probably be exposed in
the place of his master's exile, accompanied with such
strong advice that he should retrace his steps home-
ward, that his resolution to share his master's fortunes
was utterly shaken. Ludwig hastened to his master,
and told him of all that he had heard, and of the fears
and perplexities by which he was beset. His master
frankly told Ludwig that he should be sorry to part
with him, but still more sorry if his intention to follow
his master into the interior should entangle him in
serious difficulties. He urged him, therefore, to act
upon the advice which he had received; to go home to

his family; and to believe, that, if circumstances should
at any future time allow him to resume the duties which,
for nearly three years, he had faithfully performed, his
master would gladly receive him. This generous and
considerate conduct of the Count is exactly what I should
have expected from him.   Courageous, and stedfast, and
even cheerful, in the endurance of his own trials, he
would yet have been miserable at the thought of bringing
any one else, through his influence, into the same
dangers; and, therefore, not grudgingly or reluctantly,
but with sincere and hearty goodwill, he dismissed the
servant who, he believed, loved him; and furnished him
with ample means to return home.

'I have frequently seen the servant since his return,
and heard many a fresh tale of sorrow from his lips.
The good priest at Massalani, of whom I have before
spoken, has been added, among others, to the long list
of recent prisoners; but I have not yet heard what sentence has been passed upon him.

'The various houses at which I stayed have, each and
all of them, been visited, if not permanently occupied,
by Russian troops. At Wereiki especially, where Ludwig
passed a few hours one day, thirty Cossacks were found
to have taken up their quarters in the room which
had been our usual dining-room; and it can be readily
understood, that, at their departure,—whensoever that
may be, little or nothing will be left behind. Everywhere the work of pillage and oppression goes forward;
and, though the formal sentence of confiscation of the
Count's estates have not been proclaimed, yet who can
estimate the amount of damage that has been, and still
is, wilfully and wantonly inflicted upon every species of
his property?

'Of the Count himself, I am thankful to say that I
have heard a better account than I had dared to hope
for. The place of his residence is Ufa, in the province of
Orenburg. The climate has thus far agreed with him;
and the Governor of the place appears to do everything

he can to relieve the irksomeness of his exile. But the question which I cannot help asking myself, and to which I have not yet been able to find a satisfactory answer is, Why is such a man in exile at all? If he had been really guilty of doing anything, directly or indirectly, to promote the work of revolution, is it to be supposed, that, in the present temper of the Russian Government, his life or property would not have been immediately forfeited? The mere fact that his life and property are spared is proof incontrovertible that the charges attempted to be brought against him are false; and that the Government knows them to be false. But, if false, why not admit their falsehood? Why not punish, as they deserve, the perjured witnesses that dared to slander him? Why not fully reinstate the Count in the property of which he is the rightful owner, and to the improvement of which he is ready to devote his fortune and the best energies of his noble nature? There is, indeed, one answer, and only one,—but who can call it a satisfactory answer?—to be returned to these and other like questions, namely, that the unsparing rigour of Mouravieff's government forbids the display,—I will not say, of mercy,—but of any approach to equity or fair dealing towards any Pole, whose name has been in the slightest degree associated—it matters not how wrongly—with the insurgent cause. The fate of Count Bisping is but the fate of hundreds and thousands of others who, like him, are, or were, landed proprietors in Lithuania and the adjoining provinces. The system pursued has been simply a system of indiscriminate proscription; and, even whilst these sheets are passing through the press, I observe a proclamation, lately issued by Mouravieff—and now making the circuit of every journal in Europe—in which he regards with wondrous self-complacency the work done by his hands within the last few months, and prides himself in the belief that there is no longer left, throughout the extensive districts entrusted to his charge, a single inhabitant who dares to utter any other word, or to harbour any other thought than that of entire sub- mission. It may be so. He may have so closely gagged the mouth, and so heavily oppressed the heart of Russian Poland, as to make her powerless any longer to speak or to breathe. But is this to re-estab- lish order and tranquility within her borders? As well may the physician, who ascribes to his patient a malady to which he is a stranger, and drenches him with remedies which destroy him, dream that he has dis- pelled the danger, because he has silenced the moanings of pain, or made the limbs of the strong man helpless as the limbs of an infant.

'The narrative in the foregoing pages has been pur- posely confined to the notice of those persons only with whom I was brought into personal and friendly relation. The sympathy awakened within me by their distresses, I know to be a just and lawful sympathy; and, howsoever imperfect may have been the expression of it which I have tried to give, it has been given without hesitation, because I am convinced of the truth of the grounds on which it rests. If ever man were animated with a single-hearted purpose to do his duty as a steward of God's bounties, amid a people who looked up hopefully to him for help, it is the friend with whom I passed six months and more, upon the soil on which he and his people dwelt. Day by day I witnessed his honest and consistent efforts for their welfare; day by day I knew that he was stedfast and loyal to the Emperor, to whom he and his people *alike owed sub- jection.* Yet I have lived to see him torn from the home of his fathers; and the people, whom he would fain have protected and cherished, left once more to the tender mercies of the roving Cossack, or to the grasping extortion of the Jewish trafficker. His loftiest aspira- tions have been crushed in the very prime of his youth- ful manhood; and the "sun" of his brightest hope has suddenly "gone down whilst it is yet day" (Jer. xv. 9).

This is, indeed, a sore trial;—fitted, indeed, to lead every one who bends beneath the weight of it to seek more earnestly the protection of Him, who in adversity as well as in prosperity, is our surest stay. They who lean the most trustfully upon Him, and walk in the closest obedience to His will, will find that even the pathway of tribulation leads, in His own good time, to blessing. But it would be the forfeiture of this blessing, were we to varnish over, with the gloss of a false name, the hideous oppression of the country of which we have been speaking. We dare not, therefore, dignify with the name of Government (as Mouravieff and his agents would fain do), the work of plunder, proscription, and massacre, which they have carried on ; neither dare we apply the hallowed name of Peace to the desolation which they have spread over " unhappy Poland." '*

The Lithuanians seem to be of the same race with the Samogitians, and they resemble both the Poles and the Russians. Their appearance speaks of extreme poverty. Their carts are made entirely of wood, sometimes without a single piece of iron, and even the harness of the horses is often made of the more flexible branches of trees. And I have been told that the landed proprietors have become greatly impoverished, many of them possessing only half a desatin of land, and estates of insolvent proprietors being constantly for sale; while the Government has been endeavouring to provide for them by encouraging emigration to Simbirsk and Tobolsk, by offering them settlements on crown lands in these governments.

* ' Auferre, trucidare, rapere, falsis nominibus *imperium ;* atque, ubi solitudinem faciunt, *pacem* appellant.'—Tac. Agric. c. xxx.

CHAPTER II.

THERE is little to strike the eye or arrest the attention of a traveller in passing from Poland into Lithuania, which is conterminous with it throughout its eastern boundary, though extending much further both to the north and to the south. All that is seen of forest lands from the railway in both countries is alike. The forests are not continuous in either, nor confined to any well defined localities. Nor is there anything remarkable to be seen in the one or in the other.

The surface of the country is very flat, generally sandy, intersected by vast marshes and bogs, and covered with forests abounding in bears, wolves, wild boars, and other animals, amongst which is the urus, or wild ox, but this is now rarely met with, and it seems to be diminishing both in size and strength. The most common trees are the pine, the oak, and the elm. A considerable quantity of potash and pearl ash is prepared in the forests; honey is also collected in great abundance. But the people generally are indolent. The best lands lie fallow, the hay is allowed to rot on the meadows, and whole forests are at times destroyed by fire.

My journey took me through the governments of Godno, Vilna, and Vitebsk. In the government of Grodno is the celebrated forest of Bialowiéga, which extends over the district of Bialistock, and which still serves, or did some years ago, as a refuge for the last descendants of the urus of Oriental Europe. This forest is one pre-eminently suggestive of the expression *primeval forest.* 'It looks as if no provision had been made for its management, and it had been abandoned to the uncontrolled operations of

nature. It is among those aged woods which overshadow
the source of the Narova that the urus found safety ; and
there wander also the elk and the buffalo. And amid
those scenes of primitive nature subsists a distinct population—the Ruskes, almost as savage as the animals which
surround them.'

The area of forests in the government of Grodno is
958,000 desatins, of which 557,440 desatins belong to
the crown, equivalent to 28·1 desatins of forests, or 15·7
desatins of crown forests per square verst, and to 1
desatin of forests, or 0·6 desatins of crown forests per
inhabitant. The annual fellings in the crown forests
yield 12·2 cubic feet, and the revenue is 23·1 kopecs per
desatin.

The area of forests in the government of Vilna is
1,156,000 desatins, of which 387,145 desatins belong to
the crown, equivalent to 33·9 desatins of forests, or 10·4
desatins of crown forests per square verst, and to 1·3
of forests, or 0·4 desatins of crown forest per inhabitant.
The annual fellings in the crown forests yield 15·4 cubic
feet, and the revenue is 16·4 kopecs per desatin.

Of the town of Vilna, Dr Pinkerton, who travelled
through this district in the early part of the present
century as agent of the British and Foreign Bible Society,
writes :—' The town of Vilna is built among sand hills
at the confluence of the two streams Villia and Vilika.
It was founded in 1305 by Guedemin, Grand Duke of
Lithuania, of which it afterwards became the capital.
Many remains of ancient public buildings, both in and near
the town, bear ample testimony to its former grandeur. In
1748 it was burned down ; and it is recorded that at that
time thirteen churches and synagogues, about one hundred
and forty-six shops, twenty-five palaces, and four hundred
and sixty-nine store houses, &c., &c., were reduced to
ashes. Still there are at the present time many fine

churches and public and private buildings in Vilna, and daily improvements are still making. The present number of inhabitants is about 40,000, of whom one half are Jews.

'The Jewish population seem all engaged in buying and selling; the men hawking their goods about the public places, and the women seated before their shop doors knitting stockings, and loquaciously inviting the passers by to purchase of their wares. They all dress in the fashion of the Polish Jews; the men with black silk or stiff long robes, in eastern style, girt upon them, and round caps turned up with fur, with bushy beards and long hair, shining with oil and curled; the females are somewhat more European in their attire, with rich head dresses, ornamented with mock or real gems, according to their ability. The great majority of them look dirty and ragged.

'Taking leave of my Vilna friends I directed my course towards Troki, the ancient residence of the Grand Dukes of Lithuania, which lies about twenty miles to the north-west of Vilna. We reached this district town, which is beautifully situated on the lake Bressal, in the evening. This lake has a communication with the river Villia by a canal. The ruins of an ancient ducal palace are still prominent on one of the little islands which spot the bosom of this extensive lake, whose waters are clear as crystal. Troki is said to have been founded by the Grand Duke Gusdemin in 1321. It was burned down in 1390, and ruined by the Russians in 1655. The present town is divided into three parishes. In one of the churches is an image of the Virgin, which occasionally attracts a number of pilgrims from the surrounding country. The houses are built of wood. The scenery around the lake is remarkably fine in a country like this, where the eye is so seldom relieved from the sameness of prospect—of extensive plains and woods; and the evening being fine, I greatly enjoyed it.'

Conterminous with the government of Vilna on the

north-west is the government of Kovno. In the government of Kovno the area of forests is 768,000 desatins, of which 243,765 desatins belong to the crown, equivalent to 21·9 desatins of forests, or 6·9 desatins of crown forests per square verst, and to 0·7 desatins of forests, or 0·2 desatins of crown forests per inhabitant. The annual felling in the crown forests yield 14·7 cubic feet, and the revenue is 16·9 kopecs per desatin.

Kovno, the capital, is an ancient town, finely situated for commerce at the confluence of the Villia and the Niemen, bears numerous marks of depopulation, poverty, and decay. The pavement of the market place is overgrown with grass, and many of the public buildings seem hastening to ruin. The population does not exceed four thousand, of whom a great proportion are Jews, and the rest Roman Catholics.

This statement I make on the authority of Dr Pinkerton. In an account of his journeying in this district, he writes :—' From Kovno we continued to travel down the banks of the Niemen for about twelve versts, when we changed horses at a village swarming with Jews, with whom I left a Testament and a few tracts. At Svedrick, another populous Jewish town, I conversed with the people for an hour, and gave them two Hebrew Testaments and a Hebrew tract. We prosecuted our journey all night northward till about four in the morning, when we arrived at the district town of Rossiena. With some difficulty I obtained a room at the house of a Jew, where I attempted to get a few hours rest ; but the place was so cold and so uncomfortable that I felt not at all refreshed by it.

' Rossiena is chiefly inhabited by Jews ; and is now but an insignificant looking place, though formerly the residence of a Voivod under the Poles. It looks more like a village than a town ; the houses are mere wooden huts, and most of the streets not even paved. The Samogatian population of the district of Rossiena amounts to about 90,000.

'The Bishop of Samogatia has his residence here. It was a great holiday when I was there—St. John's day— and the peasantry were flocking towards it from all quarters. The general aspect of the country from Kovni to Rossiena is level, the soil sandy and clayey by turns.

'Leaving Rossiena about midday, I took the road for the district town of Shawel. At a village called Titifian there was a fair similiar to what we had left at Rossiena. In the evening we found the roads filled with country people returning from it, among whom I was pained to observe a great number intoxicated; I even became doubtful of my safety in travelling by night through extensive and little frequented roads; yet we continued our course in the rain, along very bad roads, for about fifteen miles, and then halted about midnight at a Samogatian village, where I purposed resting a few hours, but the heat and vermin in the hut was so annoying, even though I merely stretched myself on a bare bench by the side of the wall, that I could not sleep.

'We arrived in the district town of Shawel about eight the next morning. It consists of 280 dwellings, mostly wooden huts of different dimensions, the rest of brick. I found the Jews, who form the majority of the inhabitants, extremely shy, and averse to conversation on religious subjects.   .   .   .   .   .   .

'On leaving Shawel I took the road to Telsh, another district town of Samogatia, at which I arrived about one in the morning. On approaching Telsh the level and uninteresting aspect of the country undergoes a favourable change, becoming undulating, and in some parts even hilly.

'The mayor and others in Telsh assured me of a fact which will scarcely be credited—that the princes of the family of Gedroitz are so numerous, and so poor, that some of them gain their support by cutting firewood, and carting it to the Vilna market for sale. Many of them cultivate the ground for their livelihood. So low is the princely dignity fallen in Lithuania! And as to the state of the nobility of Samogatia it is also degraded beyond conception, since in the district of Telsh there are not fewer than 700 individals of noble descent who cultivate the ground with their own hands !

'The number of houses in Telsh is 160, of which one half are inhabitated by Jews, who are mostly very indigent.'

My route led from Vilna to Vitebsk. Following upward from Dunaburg the course of the Dwina, and passing New Alexandrosky on the right bank of the Dwina and Poletka we reach the town of Poletsk, one of the oldest towns mentioned in Russian history. It was formerly the provincial town of a district of the same name, which in 1796 was united to that of Vitebsk.

'It was known to the ancients under the name of Pletiscum, and in the time of Ruric, the Russian Egbert,' says Pinkerton, 'it had its own princes, who reigned there until the time of Vladimir the Great. This prince, irritated at the refusal of Rogneda, daughter of the reigning prince Rogvalde, to marry him, laid siege to the city, took it, killed the prince and his two sons, married the princess against her will, and added the principality of Poletsk to his dominions. After this the son of Rogneda obtained this principality from his father, Vladimir, and his descendants continued to reign there for several centuries, until 1305, when Guedemin built Vilna, which became then the capital of those parts of Lithuania. But even to the time of Peter the Great the Russian Tsars continued to take the title of Prince of Poletsk. In the intervening ages it was repeatedly taken and retaken by the Russians and Poles, but it generally remained in the hands of the latter until 1772, when Catherine II. united it to her empire.'

The area of forests in the government of Vitebsk is 1,738,000, of which 380,615 desatins belong to the crown, equivalent to 44·5 desatins of forests, or 9·7 desatins of crown forests per square verst, and to 2·1 desatins of forests, or 0·5 desatins of crown forests per inhabitant. The annual fellings in the crown forests yield 11·3 cubic feet, and the revenue is 12·6 kopecs per desatin.

The town of Vitebsk stands on the banks of the Dwina and Vitspa, which flow through it. It is a very ancient town, and is said to have been known to the Scandanavians and Greeks as early as the tenth century. The country around is level, and abounds in extensive forests, and the land produces abundance of grain and excellent hemp.

In the government of Moghileff the area of forests is 1,184,000 desatins, of which 180,415 desatins belong to the crown, equivalent to 28·1 desatins of forests, or 4·2 desatins of crown forests per square verst, and to 1·3 desatins of forests, or 0·2 desatins of crown forests per inhabitant. The fellings in the crown forests yield 13·4 cubic feet, and the revenue is 18·3 kopecs per desatin.

'The province of Moghileff,' says Pinkerton, 'is about 350 versts long, and 300 broad, with a population of upwards of 900,000 souls; all are Russians, except a few Lithuanians, and a numerous tribe of Jews, who formed a settlement here during the period of the Russian civil wars, when White Russia was in the hands of the Poles. It abounds in extensive forests of fir and hazel-wood; those in the district of Tchernigov are considered the finest. Numbers of the peasantry are employed in felling and floating timber to Riga by the Dwina, and to the Black Sea by the Dnieper, and to those parts of Little Russia which are but scantily supplied with this necessary article.

'The principal commodity of trade in Moghileff is leather; there are upwards of twenty tanneries; but its merchants carry on a considerable commerce also with the ports of Danzig, Memel, and Riga, in potash, hemp, leather, and grain.

'In this district is situated one of the old towns, or rather cities, of Russia, interesting on account of ancient historical associations. It is the town of Pleshkoff, built on the banks of the Velikaia and the Plescova, by the famous Russian Princess Olga, who was born in a village about eight miles distant from it, named Sibout, whence the young prince Igor espoused her for his wife. But her love for her native place was so strong, that after she had become a Christian, about the year 965, she came from Kief to propagate the Christian religion among its heathen inhabitants : and at that time, as the legend goes, founded the city on the spot indicated to her by a supernatural light from heaven, which descended upon a certain place on the banks of the two above mentioned streams. She began by erecting a church to the Holy Trinity, and the city rose around it, and became distinguished for its power and commercial importance through many succeeding ages. The evidence of this is seen in the extent of the ruins of the massive buildings required by the necessities of a crowded city of enterprising merchants, enriched by the gains of an extensive and successful trade, now yielding materials which have been employed extensively in the repairs and erection of buildings required by the demands of modern times. It is described as being in a most dilapidated condition. It is still divided into three parts—the Kremlin, the central city, and the great city. The towers and fortifications of the outer walls occupy a circle of about seven versts : these are all of limestone ; and are so reduced by the effects of the frosts in winter, and the heat in summer, that certainly a few ages more will cover the mouldering heaps with green turf.

'The population is about 10,000 people.

'When the famous Vladimir the First divided his kingdom among his ten children, whom he had by an equal number of wives, Pleshkoff fell to the lot of Sondislav, who became in consequence its first sovereign about 1030. But the fate, and even the government of Pleshkoff, was in general intimately connected with that of its rival, and elder sister, Novgorod, until its union

with the Hanseatic towns, about the beginning of the
fifteenth century, at which time they began to coin silver
money at Pleshkoff. This coin had the head of an ox,
with a crown below it ; and on the reverse, the value
marked. Republican principles prevailed in Pleshkoff
till it was subdued by the Grand Duke Ivan Vasillivitch,
in 1509. The Russian chronologer, Nestor, says that the
Christian religion was propogated as early in the region
of Pleshkoff as in Novgorod by St. Joachim the
Chersonite.'

Conterminous with the government of Moghileff, and
occupying the centre of Lithuania, is the government of
Minsk, conterminous with Vilna and Grodno on the west.
'Minsk is an ancient Lithuanian town, situated on the
river Swistoche ; sometimes subjected to the principality
of Poletsk, and at others to that of Smolensk. So early
as 1066 the two sons of the Grand Duke Jarosloff
beseiged it, took it, massacred all the males, and distributed the women and children as slaves to those who
accompanied them. In 1104 a Russian prince named
Gleb Vseslavitch reigned here, who afterwards became
Prince of Poletsk. Its fortunes were for many ages
united to those of Poland (of a Palatinate of which
kingdom it formed the capital), but was taken by the
Russians in 1656. It is now the seat of government
for the province of the same name, and of a Russian
archbishop, who takes the title of Archbishop of Minsk
and Lithuania. There is also a Russian Catholic Bishop
of Minsk. The Jews form two-thirds of the population.
At the last census there proved to be 8000 Jews, and
only 4000 Christians of all denominations, in the town of
Minsk. Many of the public buildings are of brick ; but
the houses of the inhabitants are chiefly of wood. There
are a number of manufactories of hats in this place,
which are thence exported to every part of the dominion
of Russia. The country around Minsk is fertile in grain
and pasturage. Extensive forests of pine still cover a great

part of the province, the felling and transporting of which, down the rivers to Cherson in the east, and Konigsberg in the north, form a principal branch of the industry of its inhabitants. They are reckoned at two millions of souls; and consist of Lithuanians, Poles, Jews, and Russians.'

At Minsk Dr Pinkerton met a number of distinguished Poles at the table of one of the nobles, and he remarks:— 'The spirit of national rivalry and enmity betwixt the Poles and the Russians is still strongly cherished in secret, and manifests itself in a variety of ways; the Polish ladies, for instance, I was told, refuse to dance with Poles who have become officers in the Russian regiments; and this deep-rooted national enmity against the Russians is daily nourished in the family circles and private associations of the Poles.'

In the government of Minsk the area of forests is 3,676,000 desatins, of which 1,019,522 desatins belong to the crown, equivalent to 42·4 desatins of forests, or 11·7 desatins of crown forests per square verst, and to 3 desatins of forests, or 0·8 desatins of crown forests per inhabitant. The annual fellings in the crown forests are 7·1 cubic feet, and the revenue is 9·8 kopecs per desatin.

In the government of Volhinia, conterminous on the north with the governments of Grodno, Minsk, and Moghileff, the area of forests is 2,733,000 desatins, of which 674,639 desatins belong to the crown, equivalent to 43·3 desatins of forests, or 10·7 desatins of crown forests per square verst, and to 1·6 desatins of forest, or 0·4 desatins of crown forests, per inhabitant. The annual fellings in the crown forests yield 16·9 cubic feet, and the revenue is 16·9 kopecs per desatin.

Conterminous with the government of Volhinia on the south is the government of Podolia, in which the area of forests is 589,000 desatins, of which 113,195 desatins belong to the crown, equivalent to 16 desatins of forests, or ·3 desatins of crown forests, per square verst, and to 0·3

desatins of forests, or 0·05 desatins of crown forests, per
inhabitant. The annual fellings in the crown forests
yield 53·1 cubic feet, and the revenue is 147·8 kopecs per
desatin.

Like the level of the Theiss in Hungary, the soil of
Podolen is composed of the sediment of a recent ocean,
in which a large proportion of vegetable sediment substance preponderates, and which, being saturated with
salts, needs no artifical manure to enable it to produce a
succession of the richest crops. The forests are extensive,
affording for export timber, pitch, tar, rosin, and potash,
and what is known as Polish cochineal.

A great extent of Lithuania is known as White Russia,
while the remainder of it, along with the government of
Podolia, which is conterminous with the south of the
government of Volhinia, is known as Black Russia, and
to the country comprising both is given the designation
West Russia, while to the Ukraine, comprising the governments of Tchernigov, Kiev, Poltava, and Kharkov, is
designated Little Russia, and to the Emperor is given
the designation Tsar of all the Russias.

Most of the roads in White Russia run in straight
lines, ditched and planted with rows of birch trees.
The clay cast up from each side forms the road. In dry
weather they are pleasant to travel on, but liable to be
very dusty, and after rains they become almost impassable.
And the light colour of the clay, in a state of dust, may
have confirmed the name of White Russia given to the
district; but I consider it more probable that the designations Black and White Russia were given in consequence
of the dark hue given to some districts by abounding pine
trees and oaks, and the white colouring given to other
districts by the bark of abounding birches.

' In most of the towns which I visited in White Russia,'
says Dr Pinkerton, writing of journeyings more than fifty
years ago, but there changes take place but slowly, ' a

comfortable inn is not to be found. The traveller there, as in Poland, and in the interior of Russia, must make up his mind to encounter every privation in this respect, and be glad if, on the road between provincial towns, he can procure anything better than black bread, milk, eggs, beer, and common brandy. On this account the country nobility generally carry their provisions with them when on a journey, together with a cooking apparatus, and even bedding; between the government towns they provide a sufficient quantity of white bread, poultry, wine, &c., to supply them till they reach the next town.'

CHAPTER III.

FORESTS ON THE DNIEPER.

It is not from anything peculiar to the forest management of Lithuania that it has been made the subject of
this treatise, but, as has been stated, solely because it
happens to be an integral part of the Russian empire
adjacent to Poland, which has been treated of in the
preceding part of this volume, and it supplies an opportunity of bringing under consideration the forest administration of the Imperial domains. To show this administration and management in action, I find it necessary to
take my readers beyond this limited field; and this I can
do without breaking violently away from it.

The largest river traversing Lithuania is the Dnieper,
a river extensively utilised for the floatage of wood.
By Professor Shavranoff, of the St. Petersburg School of
Forestry, there was supplied to me a series of papers by
Mr M. O. Polytaief, On Forests of the Dnieper, which
supplies the illustrations of this of which I was desirous;
and Mr Richard Sevier, of St. Petersburg, a friend whom
I found ready to give me every assistance in my enterprise which I could desire, supplied me with a translation
of the whole, which I shall insert entire.

The Dnieper, I may premise, rises in the southern
portion of the government of Tver, passes by Smolensk,
shortly after which it enters Lithuania, passes Moghileff,
and flowing thence towards Kiev, it afterwards passes
Ekaterinoslav and Cherson, falling into the Black Sea a
little beyond. It is navigable from a little above
Smolensk to the mouth. From Smolensk its course is
south-west, till it reaches Orcha, in the government of

Moghileff, under the parallel of 54° 32', whence it has a southern course through the government of Moghileff, which it divides on the south-west from that of Minsk. In this part of its course it receives numerous tributary streams; amongst others, the Dneitz, or Drouts, and the Berezina, which is united to the Dwina by means of a canal, and the Verditch, all on the right bank. At the point where it leaves the government of Moghileff, and turning to the west and south forms the boundary between parts of the governments of Minsk and Tchernigov, it receives the Soj on its left bank. Before it reaches Kiev, past which it flows, it receives the Pripett, which the Muchavice and Orginski canals connect with the Vistula and the Niemen on the right bank. After passing Kiev it receives on the same bank the Teterev, Zdvij, and Irpen. Further on it receives the Roso ; and, after receiving others on the right bank and the left, it is joined by the Bug, which traverses Podolia. By itself and its tributaries the Dnieper supplies abundant facilities for the transport of timber and firewood to the south.

In conjunction with the Bug, 'it forms a large *liman*, or swampy lagune, called Dnieprovskoi, nearly 50 miles long, and from 1 to 6 miles broad, by which it discharges itself into the North Gulf of the Black Sea. This liman extends from Cherson to Oezakoff, and in summer has hardly six feet of water. The Dnieper, seen from Cherson, resembles a vast lake studded with islands. The entire length of the Dnieper, measured by its windings, is about 1,200 miles ; in a straight line it is about 650 miles from its source to its mouth. Its depth of water at Smolensk is from 16 to 20 feet ; at Kiev, 20 feet ; at Krementchug, 20 feet ; at the rapids, 8 feet ; below the rapids, from 7 to 12 feet ; at its embouchure, from 5 to 6 feet. Its average width is estimated at 700 paces, and the surface which this river and its tributaries drain is exceeded only among European streams by that of the Danube. The Dnieper flows for the most part between high banks, the greatest elevation of which is along the east side.   The upper part of its course is through a marshy country ; in the middle and lower course it passes over numerous rocks, and between banks of the older calcareous formation.   It is broader, deeper, and more rapid than the Don, and is navigable from Smolensk to Kiev ; but below the latter town, the river is traversed by a granitic chain, and the navigation is interrupted for about 40 miles by 13 rapids called *poroge,* and also by huge blocks of stone.   Here the river presents a most magnificent sight, careering along in a bed at least 1000 feet wide, which for miles on miles is one continued sheet of boiling foam.   This space is passable for vessels of small draught during the spring floods only, and even then only with great difficulty and danger.   Works have been undertaken at various periods to render this part of the river navigable.   All merchandise for Cherson, on the Black Sea, used to be unladen at Old Samara, whence it was conveyed by land to Alexandrofsk, at the mouth of the Moscofska, a distance of about 46 miles by land.   From this spot to the mouth of the Dnieper, a distance of 260 miles, the navigation is unimpeded.   The goods that descend the rapids of the Dnieper consist almost exclusively of timber, firewood, and iron from Siberia.   Tar is also brought in immense quantities from the Polish forests.   Below the cataracts, and as far as the liman of this river, upwards of 70 islands occur, amidst which moving sands impede the navigation during summer.   These islands produce a grape called *biroussa,* which resembles the currants of Corinth. They are reported to swarm with serpents, and abound in a sort of wild cat.   Flowing through more than 9 degrees of latitude, great diversity of climate is experienced along the course of this river.   At Smolensk the waters freeze in November, and continue ice-bound until April ; at Kiev they are generally frozen from January to March.   The direction of its course from north to south delays its rise till late in the spring, as the streams which feed it from the north do not thaw till the end of April.   The

Dnieper abounds in fish, particularly sturgeon, carp, pike, and shad. This river is the *Borysthenes* of the Greeks, and the *Danapris* of the middle ages. It is first mentioned by Herodotus. Except the more southerly parts, its banks have long been inhabited by races of Slavonian origin. Towards the mouth, from the Ross on the right, and the Vorska and Soula on the left bank, the country was for a long time a mere steppe, where nomadic tribes fed their numerous flocks. By treaty with Turkey, and since the partition of Poland, both banks of the Dnieper have become the property of Russia. The lower part of its course has been the scene of many sanguinary conflicts between the Turks and Russians; the upper part, particularly the neighbourhood of Smolensk, was the scene of some severe conflicts in Bonaparte's retreat in November 1812.'

Of the forests on the Dnieper, M. Polytaief writes:—
' Under the name of Dnieper forests I understand the forests that are on the Dnieper, and also those on its affluents, therefore into the number of Dnieper forests enter those in the governments of Minsk, Moghileff, Smolensk, Orel, Koursk, Tchernigov, Kiev, Volhinia, Cherson, Ekatherinoslaff, and Poltava. In inspecting the forests of these governments I shall principally pay attention to the crown forests—or more correctly speaking, the forests under government management.
' The extent of the crown forests in the above-named governments equals about 3,000,000* desatins, and although, according to their extent, they form less than 1-30th of the whole extent of the crown forests, nevertheless, by their value in the wood trade, they are very important. Supplying timber and wood materials for exportation, the Dnieper forests have at the same time great importance in the inland wood trade, because they furnish timber to about ten governments poor in wood.

---

* A desatin is equal to 2·69972 imperial acres ; a rouble, 3s 4d, but according to present rate of exchange, 2s.

K

Constituting less than 1-30th of the whole extent of the
crown forests of European Russia, the forests under consideration realise a revenue of about 500,000 roubles,
that is one-eighth of the whole income derived from
forests.   The necessity of more assiduously guarding these
forests is the reason why in the Dnieper forests there is
about one-quarter of the whole number of foresters, and
that the expense of the local management forms about
one-seventh of all the expenses for forest management
in all the governments.   By these ciphers is sufficiently
demonstrated the importance of the Dnieper forests, as
well in the wood trade as in the income and expenditure
of the forest management.

'Of the Dnieper forests the Smolensk woods have
least importance in the timber trade, because they are
not conveniently situated as regards sales.

' Though situated at the sources of three basins, they are
for the most part distant more than ten versts from floatable
rivers, so that the expense of land transport to these constitutes the greatest part part of the outlay.   Out of 136,000
desatins only 60,000 desatins are at a distance of ten
versts from floatable rivers; the remaining mass of the forests
is much further.   Of the above 60,000 desatins, near the
affluents of the river Moskva there are not more than 1500
desatins ; near the tributary streams of the river Ourga
there are about 3000 desatins, about 15,000 desatins are
near the Dnieper and its affluents ; and the remaining
40,000 desatins are on the tributary streams of the
western Dwina and Volga   Of these floating ways—the
Volga, Dnieper, and Dwina—to the principal marts is
very far ; and only by the river Moskva to the principal
marts is it about 400 versts.   There are inconveniences of
another kind ; of the whole surface of forest estates of the
government of Smolensk about 20,000 desatins compose
about 400 separate forest parts, almost separate estates,
of which there is not one of more than 20 desatins, and
many of only 2 or 3 desatins and less.   These estates have
no importance in the timber trade, but in an economical

sense they may be rather considered inconvenient than convenient areas, as it is difficult to manage when, with a preponderance of the coniferous trees, the whole estate occupies a space of only 2 or 3 desatins. No dealer will buy on such estates, and the conservation of them entails expense.

'These surfaces, considered part of the whole forest area, give an incorrect idea of the amount of income per desatin of forest surface,—that is surface convenient for forest management, not only by the productiveness of the soil, but likewise by the extent of surface. For forest management an estate of 500 square fathoms, or 2 or 3 desatins, in such a place as the government of Smolensk, cannot be convenient. With a preponderance of coniferous trees in such a surface it is difficult to state the period of recurring fellings; and with this kind of economy one cannot expect sales, nor can one expect any income remunerating the expense of guarding, when hundreds of such estates are under one forester, and are strewn over a surface of some hundreds of square versts. The conversion of these estates into rent paying estates, or the exclusion of them from the forest area, is the urgent requirement of the forest management of the Smolensk government. In consequence of this, 20,000 desatins (which make about one-sixth of the Smolensk forests), we shall leave out of consideration, and only take into account the remaining 110,000 desatins.

'The government estates are inconveniently placed even with regard to the future railways, as well of that Vitebsk-Smolensk as of Orel-Smolensk; but both the railways pass through private forests abounding with firewood. These roads will pass near only three crown estates; but even then not nearer than 25 versts.

'But notwithstanding these unfavourable circumstances, the price of government wood is rising from year to year, the revenue from the forests is increasing, and in the course of six years it has more than doubled, from the sale of wood materials from the crown estates alone.

There was received in 1861,   -   -   -   4,605 roubles.
    „   1862,   -   -   -   6,895   „
    „   1863,   -   -   -   6,567   „
    „   1864,   -   -   -   7,032   „
    „   1866,   -   -   -   10,182   „

'In 1866 the average price of a desatin was in the Porctihie forest circuit, 18 roubles; in Dorogobouje, 100 roubles; in Donhovchina, 19 roubles; in Gjatsk, 89 roubles; in Bielsk, 19 roubles: the average for the whole governments, 49˙ roubles. In the statistical description of the government of Smolensk with regard to forests by Herne, it is seen that in 1861 the˙ average price was 32 roubles, consequently the demand and price of crown woods increase with every year, and as to the quality of the sold estates, to what kind of forest materials the increase of the revenue is due. Firewood and building materials of small dimensions, offered for sale, in the majority of cases had no buyers; the increase of the revenue was solely owing to a growing demand for forest materials of large dimensions, in consequence of fewer offers of the same from private estates. Firewood, and forest materials of small dimensions, are bought on private estates; and by an increased felling of these, the owners strive to make up the deficiency of income from their estates which proceeds from the assortments sold being of lower quality. The offers of firewood and small building materials on private estates is so considerable that the crown forests cannot compete with them, the more so that for 135,000 desatins of crown forests there are 1,526,000 desatins of private forests: that is, for 1 desatin of crown forests there are 11 desatins of private forests; or, taking 15 fathoms of firewood to 1 desatin, we receive 165 fathoms from private estates to 15 fathoms for crown. The competition is unfavourable. The constant rising of prices, and the demand for big timber, show the diminution of these latter in the private estates ; and as such materials require a recurrence of fellings every 100 or 150 years, we think that the importance of the crown forest estates in the government of

Smolensk will increase every year, if in the majority of these be introduced a forest economy with distant recurrence of fellings. And no objection can be raised to distant recurrence of fellings, as in most of the estates the plantations and soil are quite suitable for such a form of economy.

'With regard to the value of a desatin of forests, the following estates are worthy of notice :—

' *Klevtzovokholmovskaia.*—This estate forms part of the Gjatsk forest circuit, and is in the Gjatsk district ; it measures 859 desatins, 473 fathoms, of which 776 desatins, 1109 fathoms, are covered with wood ; 8 desatins by buildings ; and 74 desatins, 1764 fathoms, are waste. The forest surface on 407 desatins pine predominates ; on 300 desatins, fir ; the average age of the plantation is about 100 years ; 68 desatins are taken up by cuttings and glades. Out of this estate, in 1864, were sold 10 desatins for 1115 roubles, that is 111 roubles the desatin ; in 1865, 10 desatins for 1273 roubles, or 127 roubles the desatin ; in 1866, 10 desatins for 890 roubles, or 89 roubles the desatin. This decrease was occasioned by the dealers ; and in consequence of it in 1866 they, by the proposal of the taxator, began to leave seed trees —60 pine trees to a desatin, though the pine trees make the principal value in a clearing. Having grown in a dense forest, these trees have their summit at a height of 6 or 7 fathoms from the earth, and an even trunk without branches. Being in a plantation where, to a desatin, there were about 500 trees, the pine trees left for bearing seed will be blown down before they bring any profit by sowing the clearing. This estate has a great similarity to the Zabolotsky estate in the government of Moghileff. In the Zabolotsky estate it was likewise proposed to leave seed-bearing trees in a similar plantation, and the result was only the reduction of its value, and not the sowing of seed. In such plantations the natural sowing of the clearing, with indiscrimate sale, can only be aided by arranging the clearing, so that the felling of one

stripe shall not follow that of another proximate
to it within five years, or generally not earlier
than the lapse of that number of years, after which seed
years occur.

'With the small breadth of the clearings the
sowing of them will take place from the surrounding
plantations. In planing out the clearings, one by the
side of another, for five, seven, or ten years, as this is
sometimes done in order to diminish the labour, the
leaving of seed trees will not assist reproduction. The
seed trees will be broken by the wind, and the leaving
them, without bringing any assistance to reproduction,
will only bring a loss in the sale.

'*Holovin Waste, Gjatsk Forest Circuit.*—Surface 371 desatins, 257 fathoms ; of this number 367 desatins, 257 fathoms
are under forest ; 4 desatins, waste land ; and 25 desatins,
330 fathoms, under cutting and glades. Of the space
occupied by forests, 312 desatins are under fir, and 3
under birch. The ripe plantations are about 70 per cent. ;
those of middle growth, 30 per cent. Of this estate there
were sold in 1864, 7 desatins for 382 roubles, or 54 roubles
per desatin ; in 1865, 5 desatins for 275 roubles, or 55 roubles
per desatin ; in 1866, 5 desatins for 278 roubles, or 55
roubles per desatin.

'*Senskaia Estate in Dorogobuj Forest Circuit.*—It is
generally divided into crown forests, with the square contents of each, shown separately, but as in an administrative point of view, it forms one whole, and we show the
items. The general surface is 2,318 desatins, 654 fathoms.
Of this space the forest surface is 1,845 desatins, 1,354
fathoms, of which 170 desatins are under pine ; 1,502
desatins under fir ; 70 desatins under birch, alder, and
aspen ; under clearings, 103 desatins ; under buildings,
17 desatins, 400 fathoms ; waste ground, 455 desatins,
300 fathoms. From this estate there were sold in 1865, 24
desatins for 1,912 roubles, or 79 roubles per desatin ; in
1866, 21 desatins for 2,205 roubles, or 105 roubles per
desatin.

'To the number of remarkable estates of the Smolensk government with regard to value of clearings may be added :—
'*Orel Estate, District Eln, Dorogobuj Forest District.*—It occupies about 306 desatins, 2,017 fathoms, of which 286 desatins, 2376 fathoms, are forest surface; and 19 desatins, 2,041 fathoms, waste; and 26 desatins, 1260 fathoms, under cuttings and glades.   Of the forest surface 12 desatins are of pine ; 214 desatins of fir ; and 15 desatins of mixed. With regard to age, young plantations are 5 per cent., and ripe 95 per cent. In 1864 there were sold 4 desatins for 300 roubles, or 75 roubles per desatin; in 1865, 3 desatins for 217 roubles, or 72 roubles per desatin ; in 1866, 3 desatins for 218 roubles, or 72 roubles per desatin.

'All these estates supply wood materials of large dimensions, and the sale of their yearly cuttings goes on well, whereas cuttings of small building materials and firewood seldom find buyers ; therefore, we repeat the gradually rising importance in the wood trade will effect those forests in the government of Smolensk which have a store of large dimensions ; and consequently for the greatest number of estates there is required forest management with a remote return of cutting.

'The importance of the Smolensk-Dnieper forests in local trade is unimportant, and one can calculate only on the trade beyond the frontier of the government, and on three markets, the mouth of the Dnieper, Riga, and Moscow.   The Moscow market, from its proximity, we consider the principal market for the Smolensk-Dnieper forests.   In consequence of this connection of the Smolensk forests with the Moscow markets, we consider it not out of place to make some comparison of the taxes of the Moscow government with the taxes of the government of Smolensk.   The greatest similarity is between the districts of Mojaisk and Volokamsk of the government of Moscow, and the Gjatsk district of the government of Smolensk.   The difference between them, in a commercial point of view, consists in this—that in the

Gjatsk district the local demand for material is much less than in that of Mojaisk and Volokamsk ; and with regard to float-age it must be remarked that, by the course of the rivers Konsa and Moscow, the district of Gjatsk is 100 versts further from Moscow. The medium price of birch wood in the districts of Mojaisk and Volokamsk is 4 roubles for highest quality, and in the district of Gjatsk, 4 roubles; and wood of lowest quality is 2·10 kopecs in all, though the distances from the principal markets are different.'

Of forests in the government of Orel he writes :—
'The forests of the Orel government, like the forests of Smolensk, are situated likewise on the basins of the Dnieper, Oka, and Don, but the crown forests of the Orel government have a much higher importance in the wood trade than those of Smolensk ; the Orel crown forests form a much larger extent of the whole surface under forests than those of Smolensk, and, with regard to communication, are under still under more favourable conditions.

'The whole extent of crown forest estates which are under government management is distributed in the districts as follows :—

			Of this number the	
In the district of Orel,	2,458 des.	forest surface	2,297 des.
,, Mtzensk,	502	,,	502
,, Karatchev,	25,142	,,	19,813
,, Briansk,	161,640	,,	131,078
,, Tronbtcheosk,	87,890	,,	72,544
,, Sevsk,	10,710	,,	9,036
,, Dmitrovsk,	770	,,	722
,, Kromin,	1,995	,,	1,902
,, Liven,	7,645	,,	5,600
,, Jeletz,	15,703	,,	13,000
,, Bolchov,	132	,,	132
,, Malo-Archan-gelsk,	1,945	,,	1,845
	344,532 des.		255,533 des.

'According to approximate information, by which the local forest administration are guided, there are of private forests :—

In the district of Orel,		14,125 desatins.	
,,	Mtzensk,	10,747	,,
,,	Karatchev,	36,566	,,
,,	Briansk,	234,897	,,
,,	Tronbteheosk,	214,525	,,
,,	Seosk,	51,125	,,
,,	Dmitrovsk,	37,249	,,
,,	Kromin,	7,840	,,
,,	Liven,	10,048	,,
,,	Jeletz,	34,471	,,
,,	Bolchov, about	24,000	,,
,,	Malo-Archangelsk,	24,000	,,

699,597 desatins.

'The information anent the crown forests is much more correct than the information as to private forests, particularly in the districts of Bolchov and Malo-Archangelsk. If, instead of the general extent of crown forests, we take into account only the forest surface, which even, with regard to private estates, is shown after excluding all the appurtances contained in the limits of the forest estates, then we will perceive that the crown forests are almost two and a-half times less than the private forests. Taking the districts there are important deviations—for instance, in the Mtzensk district the crown forests are only one-twentieth part of the private forests, and only in the districts of Karatchev and Briansk are the crown forests more than half. With such proportion the Orel crown forests may have a much higher importance in the forest trade than those of Smolensk.

'Taking the number of inhabitants of the district as a foundation census of 1862, we find that—

		With a population of	For 1 man there is general surface.	Of which extent forest.
In the district of	Bolchov,	121,113	1·94 dec.	0-012 dec.
,,	Orel,	162,981	1·10	0-094
,,	Mtzensk,	90,051	2·32	0·115
,,	Karatchev,	98,822	3·42	0·57
,,	Briansk, ⬦	110,168	6¡23	3·2
,,	Tronbteheosk,	95,473	5·63	2·87
,,	Seosk,	112,174	2·82	0·53
,,	Dmitrovsk,	82,325	2·54	0·448
,,	Kromin,	96,128	1·85	0·099
,,	Malo-Archangelsk,	145,797	2·02	0·673
,,	Liven,	224,317	2·25	0·06
,,	Jeletz,	205,432	2·41	0·22

There is for each man—

		Crown forests.	Private forests.
In the district of	Bolchov,	0·001 dec.	0·11 dec.
,,	Orel,	0·014	0·08
,,	Mtzensk,	6·005	0·11
,,	Karatchev,	0·2	0·37
,,	Briansk,	1·1	2·1
,,	Tronbteheosk,	0·22	2·5
,,	Seosk,	0·08	0·45
,,	Dmitrovsk,	0·008	0.44
,,	Kromin,	0·019	0·08
,,	Malo-Archangelsk,	0·013	0·016
,·	Liven,	0·02	0·04
,,	Jeletz,	0·06	0·16

'Consequently with regard to the abundance of wood, the government of Orel can be divided into three groups— the districts of the basins of the Dnieper, the Don, and the Oka.

'To the first group belong the districts of Briansk, Tronbteheosk, Seosk, and Karatchev. To it refer likewise the greatest part of the crown forests, viz.:—

						Of which are under forests.	
1. Briansk circuit,	61,399	desatins.	2,230	fathoms.	50,457	desatins.
2.		30,602	,,	448	,,	14,092	,,
3.		69,639	,,	1,354	,,	66,529	,,
1. Tronbteheosk,	43,246	,,	145	,,	32,315	,,
2.		38,288	,,			35,280	,,
3.		6,356	,,			4,942	,,
Karatchev,	25,142	,,			19,813	,,
Seosk (namely—						
Seosk district),	10,710	,,			9,036	,,

' In these forest circuits the greatest part of the forests constitute crown estates, whereas in the remaining forest circuits they belong to the peasants.

' The second group contains Jeletz, Liven, and Malo-Archangelsk.

' To the third group belong the districts of Orel, Bolchov, Mtzensk, Dmitrov, and Kromin.

' It is evident that the two last groups must be referred to places with little wood. The importation of forest products into these districts from the south and east is impossible in consequence of their being no forests in these parts ; the importation by the river Oka, and generally from the north—from the governments of Tula and Kaluga— also cannot be great, because the government of Tula has no excess of forest materials, and the forest materials of the government of Kaluga will always find an increasing demand on the Orel-Moscow railway, and being floated down the Oka, will find a better market in Moscow—because those landed at Serpuchov will be transported by rail only 90 versts, whereas from Serpuchov to Orel by railway is much further, and besides the floating by the Oka against the stream, and frequently against a contrary wind, is long, and costs dear. In the meantime the demand for forest materials in these districts must increase rapidly. In Orel three branches of railways meet, in consequence of which the sale of the manufactures of the government of Orel must increase, and this increase must increase the consumption of forest materials, because in the government of Orel leather and tallow fabrics, &c., have a great importance, and consume large quantities of forest materials.

'So the importation of forest materials to the places destitute of wood in the Orel government is inevitable; it will increase yearly, and the prices, with the opening of railways, will rise greatly. The most profitable supply of forest materials will be from the Dnieper forests of the Orel government, that is from the first group to which refer the districts of Briansk, Tronbteheosk, &c. This part of the Dnieper forests, as the most easterly, and is near to the woodless stripe, the principal places of which must be counted Kursk and Orel. Even now the lack of wood in this stripe calls for a constant land transport of forest materials to the neighbourhood of Orel and Kursk, and further in the direction of Woronetz and Kharkov.

'The forests of the Dnieper basin have a future. The Orel-Smolensk railway cuts through the northern part of this group of forests; the Kursk-Kiev passes not far from the southern part; the navigable river Dwina passes almost through the middle of this group, and affords a possibility of cheap conveyance of forest materials from the furthest parts of these forests to one or other of the railways, each of which leads to the important wood markets of Orel and Kursk. Besides this the Kursk-Kiev railway, passing by the left bank of the river Seyina, enables the transport of forest materials to be made to the north-west part of the government of Kharkov, and the south-west part of the government of Kursk, where the local forests are insufficient for local wants. The river Dwina, passing through the middle of this group of forests, permits of the floating to the lower governments of those sorts for which their might be no demand in the aforenamed two markets. These circumstances favouring the sale from the crown estates lying on the Dwina, in the governments of Orel and Ticherigov, confer on these estates the characteristic of one economical unit, for which even a separate administration would not be unprofitable

'An increased demand, and yearly increased revenue, being fully guaranteed to the forests of this group, their importance in the wood trade will rise with every year, because in addition to the demand of other districts of the Orel government there is a great local demand for forest materials for different industries, among which not the least important are saw-mills. With such a future for the Dnieper forests of this group there should be no hurry to sell, that is to offer for sale, such a quantity of forest materials as would exceed the demand. Such offers only call forth lower prices, and consequently less income, from a desatin of forest surface. And here it would be advantageous to have a more commercial system of sales. With that importance of wood in the wood trade which this group has it would be advantageous to offer for sale only that quantity of forest materials which had been felled in the previous year, and not the full quantity which the estate could produce for sale. Of course in that case when by the state of the property more can be felled than is annually sold, and part of which every year remains unsold, it is the more necessary that there should not be offered for sale the surplus left unsold from former years. The offer for sale of any goods in a greater quantity than there is demand for is unprofitable for any economy, and particularly in cases when an increased demand in a short time, and increased prices for the goods, is foreseen.

'The commercial system of sales in these cases appear to offer more advantages to the wood trade than the present. This commercial system is not to offer for sale more than the quantity sold in the preceding year; and if in any current year a good deal more be in demand than in the preceding year, then in the next (that is the third year) to offer for sale a quantity only so much more that none should be left unsold. With such a commercial system of sale the offer for sale will correspond with the demand, and this is the principal thing in every traffic; as for the prices, they will rise every year in conformity with the demand, and therefore the highest price per desatin will be received.

Means of Conveying Timber from the Forest Districts in the Government of Orel.

' The transport of forest materials from the forest belt to that destitute of wood was carried on before the emancipation of the serfs by the proprietors, by cartage ; grain was brought to the wood belt, and timber taken back. The proprietors derived the principal benefit from this ; and it may be said that this cartage was concentrated in the hands of large capitalists. With the freedom of the serfs the character of the transportation changed ; it fell into the hands of small peasant proprietors, because extensive areas of private forests were sold to commission traders for ready money; and the wish to get a quick return of capital led to sale in retail on the estates. The peasantry, accustomed to cartage, began to buy different sorts of wood which they conveyed to the destitute places, on making a previous agreement as to price and quality with the persons requiring timber. In the course of the winter they could return several times to Orel and bring forest material instead of the grain which they took. The possibility of getting on in these cases with a small capital then concentrated the wood trade in the hands of small dealers. Now a third plan commences with the system of conveying wood materials.

' The railroad, although it affords a means of supplying Orel with wood material cheaper than was done by cartage, will require a much larger capital ; and therefore the transportation, and with it the wood trade, will again concentrate itself in the hands of extensive wood dealers. The remark may be made that formerly the conveyance of forest materials may have been partly met by the payment for the cartage of grain ; but now the grain will also go by railway, and cartage will only be required to take wood to the railway, fewer carts will be required, and only for a short time. The waggons that take the grain will be loaded principally with timber on their return; and at the railway stations along the line woodyards will be opened, from which only the cartage will commence, and with this a small wood trade, so that the railway having increased the demand for forest materials, will considerably diminish the expense of its transport, and this circumstance is particularly favourable to forest proprietors. But in any case cartage will not loose its importance in the wood trade in the zone destitute of wood of the Orel and Kursk governments, and in making out the estimates this must be taken into consideration.

'With regard to Orel itself the railway and its tariff alone will be important, but with regard to places lying between railways special attention must be paid to cartage, because it will materially influence the prices of wood materials.

'The cost of land transport, although subject to different changes, has, however, sufficiently defined foundations—the cost of food, and consequently the degree of fertility of the year, the distance of the forests from the places of consumption, and the state of the weather, as influencing the duration of the transport; these are the points on which generally the prices per pood of the land transport depend. The principal reasons for differences in the cost of transport of different kinds of goods are the cost of packing and the degree of injury goods are liable to in transportation. These same circumstances determine the cost of transport of forest materials. The quantity of poods of one or other sorts of forest materials determine the quantity of horses required for cartage; therefore the weight of the beam, spar, or other sort of forest material, must be taken into consideration in stating the expense of transport, the quantity of each sort of materials that can be placed on each one horse cart, and likewise the cost of a one horse cart for a given number of versts, or for the time necessary for going the distance. In transportation of forest materials of more minute workmanship, such as rakes, a more careful packing is required, and more care must be taken during transportation. The products of wood technically prepared particularly require more costly packing, and therefore their transport is much more expensive than other forest materials, particularly such as require the least care in transport, and therefore the land carriage per pood is cheaper. On these considerations we think that previous to making an estimate we must ascertain the weight of each sort of forest materials, and the number of horse carts required for the transport of this assortment, if it be of large dimensions; and if of small dimensions, then in the number of pieces of this assortment that can be placed on one cart. The chief foundations for such estimates are the following:—The weight of a cubic foot of wood, the cubic contents of one or other assortment of materials, and the normal strength of a peasant's horse. These are of course known, but hitherto have scarcely been taken into account in estimates. In making the following table we have had in view to give a foundation for such calculations requisite in each case. With regard to the cubic contents (shown lower in the table), in most cases the medium for the three dimensions should be increased a little if it can be foreseen that an unequal number of different dimensions, and principally great, will be received; in the contrary case they should be lessened.

'With this calculation the weight of a cubic foot for coniferous and soft leaved kinds in a half-dry state has been considered 40 lb., and for the hard leaved sort, in the half-dry state, at 60 lb; for soaked timber (floated) on landing it a quarter more should be added to the weight.

Length in Archines.	Thickness in Vershocks.	Average Cubic Contents in Feet.	Weight. Coniferous and Soft Leaved.	Weight. Hard Leaved. (Poods)	Number of Carts For each Hundred. For Coniferous Timber.	For Leaved Timber.
STAKES.						
3	1·2	40 in a hundred	40	60	2 Carts, say 50 pieces per Horse.	3 Carts.
4½	1·2	70 ,,	70	105	4 ,, 25 ,,	5 ,, of 20 per Horse.
6	·2	100 ,,	100	315	5 ,, ,,	15 ,, ,,
POLES.					PER HUNDRED.	
9	1·2	210	210	315	10½	15
12	1·2	330 ,,	330	495	16¼	25
15	1·2	480 ,,	480	720	24	36
BOARDS.						
9	2·3	350	350	525	17½	23
12	2·3	550 ,,	550	825	27½	41
15	2·3	750 ,,	750	1125	37½	56
BEAMS.						
8, 9, 10	4, 5, 6	13 per Beam.	13	19½	2 Beams per Horse.	1 Beam per Horse.
	7, 8	23 ,,	23	34½	1 ,,	2 Horses.
	9, 10	36 ,,	36	54	2 Horses.	2 ,,
	11, 12	51 ,,	51	76	2 ,,	3 ,,
	13, 14	74 ,,	74	111	3 ,,	5 ,,
11, 12, 13	4, 5, 6	17 ,,	17	26	1 Beam per Horse.	1 ,,
	7, 8	27 ,,	27	40	1 Horse.	2 ,,
	9, 10	43 ,,	43	64	2 ,,	3 ,,

Archine = 27 inch.   Vershock = 2¾ inch.   Pood = 36 avoirdupois.

L

FORESTRY IN LITHUANIA.

Stakes. Length in Archines.	Thickness in Verschocks.	Average Cubic Contents in Feet.	Weight. Coniferous and Soft Leaved. (Poods.)	Weight. Hard Leaved. (Poods.)	Number of Carts For each Hundred. For Coniferous Timber. (3 Horses.)	For Leaved Timber. (4 Horses.)
	11, 12	64 per Beam.	64	96	3	4
	13, 14	87 ,,	87	130	4	6
14, 15, 16	4, 5	20 ,,	20	30	1	1
	6, 7	32 ,,	32	47	1½	2
	8, 9	52 ,,	52	78	2	3
	10, 11	75 ,,	75	110	3	5
	12, 13	105 ,,	105	155	5	8
17, 18, 19	4, 5	25 ,,	25	37	1	2
	6, 7	40 ,,	40	60	2	3
	8, 9	64 ,,	64	96	3	4
	10, 11	93 ,,	93	139	4	6
	12, 13	128 ,,	128	192	6	10
20, 21, 22	4, 5	31 ,,	31	46	2	2
	6, 7	48 ,,	48	72	2	3
	8, 9	78 ,,	78	117	3	6
	10, 11	114 ,,	114	171	6	9
23, 24, 25	4, 5	38 ,,	38	57	2	2
	6, 7	58 ,,	58	87	2	4
	8, 9	92 ,,	92	138	4	6
26, 27, 28	4, 5	47 ,,	47	70	2	4
	6, 7	70 ,,	70	105	3	3

(Average Cubic Contents, for 3 Beams. Number of Carts, for 3 Beams.)

'A cubic fathom of firewood, coniferous kinds, requires about ten carts, a cubic fathom of birch and hard leaved kinds will require thirteen carts. Of the loads brought to market they frequently pile a cubic fathom out of eight loads. Generally one must distinguish three kinds of cubic fathoms of wood :—Forest fathom, that is the fathom received from the labourers in the forest who are paid by the fathom ; trade fathom, that is the fathom as piled up in the woodyards for sale by the dealers themselves ; and contractors' fathom, that is the fathom piled by the labourers who receive pay for cartage by the fathom. This fathom is called contractors, because on delivering wood per contract, the carters, who likewise pile the wood in the place where it is received, receive an addition if out of the number of fathoms taken by them from the yard more come out at the place of delivery. In cubic contents the forest fathom is the largest.

'For each vershock of thickness, and archine of length, one need not make a calculation, because in practice in carting such a calculation will not be applicable. There may be changes in these ciphers—for instance, at the commencement of the day, when the horse is fresh, more is put on the load, and with a beam they put several stakes cr poles, &c., but on an average the quantities stated in the tables may be accepted.

'The number of horse working days for cartage must be determined according to the distance of each forest estate or clearing from the floating river, or the principal market at which it is expected to make the most remunerative sale. In carting to the river, if the timber be prepared at the distance of 5 versts, it is estimated that the cart will go twice ; with a distance of 1 verst it will go 10 times ; with a distance of 16 versts, once. If the wood be carted for a long distance, one may calculate on 50 versts per day by winter roads ; but in rainy, and generally bad weather, 40 versts. By the price of the horse working day is determined the price of each article of forest material. For instance, if a horse working day is worth 40 kopecs, and the timber is prepared 4 versts from the river, the cartage of a beam 9 archines by 7 vershocks, of coniferous kinds, will be 10 kopecs, because a one horse cart will bring four beams per day; but in winter, from the scarcity of work, a horse working day may be 30, even 25 kopecs, then the cartage will cost about 7 kopecs; the cartage over 10 versts will be 20 and 24 kopecs, and over 16 versts 30 and 40 kopecs. The difference is more noticed in those cases where large materials are transported, requiring several horses.

'When the distance of the principal mart is determined then the cost of a one horse cart per day is determined. In this case the taxation of wood comes into contact with the general official prices, which can be taken as a groundwork, or at least is taken into consideration.

'In transporting to great distances the price per pood for transport to a given mart then must be considered; in transport by water in ships the freight per pood; and in transport by rafts, the number of days they were on the way, and the number of workmen required for each raft. The price of working day by the general prices may give a means to determine the cost of transport.

'The distance of the estate by water from a given market can be determined with great accuracy and ease with the assistance of the charts of the General Staff and of Poltoratsky and Ilyin.*

'The third basis for forming estimates, which at present it is difficult to obtain, but which in future will be easy, are the market prices for forest materials. These prices are subject to different changes, and if these changes were reported each time one could judge pretty accurately of the value of the forest materials; but generally these prices are not noted, and in making estimates there are raked up some data from memory; and mistakes are of course

---

* The Government has published a chart of Russia by Schoubert (scale of ten versts per inch), and on the same scale a particular atlas of the Russian Empire by Poltoratsky.

unavoidable. If we take into account that the service
in one place of the government or local foresters, is not
long, then one cannot but agree that there may be often
cases in which the making of estimates for a given place
falls to the duty of one who has had no means of getting
acquainted with the conditions of the demand and sale.
In making estimates on this system where it is not known
in a given government by what they are guided in the
adjoining governments, there arise in consequence unavoidable differences in the estimates, which, if not
mutually contradictory, are not justified by the reality.
To avoid this it is desirable that the local foresters, or
forest revisors, in the course of the year, should supply to the
Journal of Forestry three or four times a year information
of the market prices of the forest materials in one or
other of the markets. If such reports on the part of
the foresters (there are 400 of them) be considered
inconvenient, then from the revisors (of whom there are
100) it appears to us it may be practicable and sufficient,
as with the help of the local foresters each forest revisor can
collect information as to the prices in those markets
which he has to visit in making the revision. This information should be supplied three to four times a year to the
editor of the journal, and the supply and placing of these
statements in the journal should be arranged in a formal
manner. If there were such a collection of information
as to market prices it would be possible to see the regular
changes, and likewise the occasional changes in prices;
and in making estimates, for one government to take into
consideration the prices in other governments. The
correctness of these prices could be timely considered and
rectified. Such information would further afford a
possibility to private forest proprietors to keep fixed
prices, so that even between them there would be more
unity in determining prices then exists at present; the
taxators would find it possible to guide themselves
by the varying prices of each locality for several years,
and persons specially learning forest economy would have

an opportunity even at school to get acquainted with present prices of produce of forest economy in different localities, which might be of use in future to forest proprietors.

'The determination of the cost of preparing each assortment is possible, if a stated number of working days necessary for preparing such were fixed; a similar, so to say task allowance, should be taken as a general groundwork for all estimates, and then the cost of a working day can be determined by the official prices.

'On these foundations estimates are determined easily and correctly. Difficulties may occur only in such localities in the governments of Volinia, Minsk, Moghilev, Kostroma, and similar governments, as the market value of forest materials is less than the cost of working days required for felling and cartage. For instance, there are many places where a fathom of wood in the market is 3 roubles—whereas in felling it two working days were employed, and 10 horse working days for carting the same, and for 12 days is received 5 roubles or 23 kopecs per day ; and whereas in the same locality a workman receives 25 to 30 kopecs per day, and a labourer with a horse 40 to 50 kopecs. In such cases it is impossible to fix the value of forest materials by the market prices. But such prices only prove that the greatest part of forest materials that appear in the market are acquired in the forest estates without the consent of the proprietors.

Remarks on the Wood Trade in the Government of Orel.

'The north-west district of the Orel government, Briansk, containing a large mass of forest as stated before, has besides, this important privilege that by the rivers Dwina and Bolva it is connected with the wooded parts of the governments of Kaluga and Smolensk. The railway in the course of construction near Briansk, and the river Dwina, connects Briansk with places where there is a scarcity of wood, by easy and cheap ways of communication for the sale of wood materials. For these reasons the district of Briansk will be the principal centre of the wood enterprise. At present there are three steam and one water power sawmills : two of them are not specially for sawing, but they can be applied to that purpose—principally they are for grinding corn. On the principal of these fabrics the sawing apparatus is for two frames each of 10 saws ; and for cutting off the edge there is a large saw. At the other mill there are 40 saws; here they make the same sorts of timber as at the first. At the third mill there are 50 saws ; and the sawmill worked by water has 44 saws in 4 frames. The total number of logs sawn in these mills is about 50,000, for the most part not longer than 3 fathoms, with a thickness of from 6 to 12 vershocks. Boards are prepared of $1\frac{1}{4}$ to 2 vershocks thick, and 7, 8, and 9 archines long ; and deals of $\frac{3}{4}$ vershocks thick. The produce is floated down the Dwina to the southern governments, or sent by cartage to the parts destitute of wood in the governments of Kursk and Orel. With the opening of the railway the produce will be sent principally by rail to Orel and Kursk. The existence of sawmills, and the exhaustion of neighbouring private forests, draws attention to the crown forests ; and the opening demand for wood leads to the conclusion that it should not be offered in greater quantity than the demand. At the forementioned sawmills the prices for boards are—

9	archines long,	and	2	vershocks thick,		60	kopecs.
8	,,	,,	2	,,	,,	50	,,
7	,,	,,	2	,,	,,	45	,,
9	,,	,,	$1\frac{1}{4}$	,,	,,	35	,,
8	,,	,,	$1\frac{1}{4}$	,,	,,	30	,,
7	,,	,,	$1\frac{1}{4}$	,,	,,	25	,,
9	,,	,,	$\frac{3}{4}$	,,	,,	18	,,
8	,,	,,	$\frac{3}{4}$	,,	,,	15	,,
7	,,	,,	$\frac{3}{4}$	,,	,,	12	,,

'The average width of these sorts is 6 vershocks ; when wider 12 kopecs is added for each vershock, so that up to

12 vershocks the addition will be 72 kopecs, that is a board
of 9 archines long, 12 vershocks broad, and 2 vershocks thick,
will be worth 132 kopecs, and if there be a demand, 150
kopecs. The floatage to Kiev for 100 boards is 3 roubles
50 kopecs, or 3½ kopecs per board; the floatage of deals
half that charge. To Krementchug the floatage is per
board, 4½ kopecs; per deal, 2¼ kopecs. On the rafts to
Kiev the labourer receives 30 roubles; on ships the
captain gets 50 roubles, and the labourer 20 roubles; to
Kherson the captain gets 80 roubles, and the labourer 30
roubles. The cost per pood is to Kiev 12 kopecs, to
Krementchug 16 kopecs, to Kherson 20 kopecs. Formerly
the prices were not so high, the demand for labourers for
railways have raised the prices; and with the completion of
these it is expected the prices will fall 3 to 5 kopecs per
pood. With the transport by land the expenses are 10 to
25 kopecs per board and deal, according to size, which
decides the weight, and consequently the number on each
cart.

'In Briansk there are 20 brickworks, and fuel costs
there about 550 kopecs; but the general quantity of wood
used in these works will barely amount to a few cubic
fathoms. Glass and crystal fabriques are very common; and
the quantity of potash brought for them from Nijni-Nov-
gorod amounts to 30,000 poods. If the wood dealers of the
governments of Moghilev, Minsk, and Volhinia, were to
direct their attention to this then the great remnants of
oak would not be removed in the clearings to their
detriment, but would bring in a pretty considerable income.
The total amount of firewood used in these fabriques is
about 75,000 cubic fathoms. In the district of Briansk
there is an iron foundry remarkable for this, that the charcoal required for it is made in stoves, and consequently is
prepared with the greatest economy, which is a rare thing
in our fabriques.

'Besides these fabriques there are sugar, brandy, and
other fabriques requiring fuel. With the completion of the
railway the activity of these fabriques will increase, and

consequently the consumption of fuel will considerably increase, and likewise the income from the crown forests on the Dwina and its confluents.

'In the Karatchev district is a cooper fabrique worthy of notice, in which they use 1000 heaps of bands for casks, and about 30,000 hoops. 1000 hoops is worth 13 roubles; a heap costs about 3 roubles, and is generally supplied by land in the districts of Klimovitch of the Moghileff government, and Roslave of the Smolensk government. The manufacture of these bands costs on the spot 175 kopecs for 100 pieces of 3 archines long; of 2¼ archines long 1 rouble for 100 pieces; 1½ archines, 80 kopecs; and 18 vershocks, 50 kopecs per 100, according to length of band, and the number used for a cask, tub, or barrel (from 10 to 26 hoops); the produce is sold from 6 roubles to 240 kopecs, and from 1½ roubles to 2 roubles. In all there are manufactured about 3000 pieces, principally tubs are prepared, which are sold on market days in the district of Orel; sometimes they get orders from brandy fabriques.

'In the district of Tronbteheosk, on a confluent of the Dwina, from the left side in the village Promniss, situated in a woody place, there is a sawmill worked by water which draws by the local prices 10,000 roubles.

'Wood materials are much used by leather fabrics of the government of Orel. These are in all parts of the government, but the principal centres are Bolchov and Jeletz, and after them the districts of Tronbteheosk, Seosk, and partly Briansk. The total amount of skins manufactured amounts to 600,000. According to this number the consumption of wood materials is determined as follows. For manufacture of skins put into one vat are required—

Bark oak,	250 poods.
Tar,	8 ,,
Pitch,	2 ,,
Firewood,	½ cubic fathoms.

The smallest number of skins put into a vat is 90; the

greatest number, 350; the medium about 200; therefore for the manufacture of 600,000 skins 3000 vats are required, and consequently

Of Bark,	-	-	-	-	-	-	-	750,000 poods.
Tar,	-	-	-	-	-	-	-	24,000 ,,
Pitch,	-	-	-	-	-	-	-	6,000 ,,
Wood,	-	-	-	-	-	-	60,000 cubic fathoms.

'The greatest influence on the forest and wood trade in this case is the destruction of the bark; and as the bark of young trees is required, or of young branches of trees from 3 to 5 years old, then the preparation of sale bark must act generally injuriously to forest economy. In this case woods are partly helped by the circumstance that willow bark is frequently preferred, although it contains less tanning acid; it pounds easily, and being more binding, it is used for manufacturing the rougher sorts of leather.

' Willow bark is brought from the neighbouring governments, and is partly collected in the government of Orel. The price of it at Bolchov and Jeletz, the centres of consumption, is 30 to 40 kopecs per pood; in winter it falls to 20 kopecs. Willow bark is often 5 kopecs higher than oak bark. The average in the districts of Seosk and Tronbtcheosk are the same. If we take 20 kopecs as the average price of the whole quantity consumed, the value of the bark will be 1,500 roubles. But as the price at Jeletz is more frequently 35 kopecs, then the amount will be about 2,000 roubles : and this is the case, because in Bolchov, to which place the bark is brought from the government of Kaluga, in consequence of the unequal supply, the price sometimes rises to 40 kopecs. Such a consumption of willow bark obliges us to direct attention to the marshy places of the crown estates, and to adapt them to forest economy, in order to receive the requisite material for the leather fabriques of the Orel government, which form a very important produce of this locality. The prices of bark from the marshy or so-called perspiring places are higher than for bark from dry places. The plantation of willow does not present any difficulties; and it can with great facility be produced in many places of little use. On the river Dwina and its confluents there can be no want of spaces fit for such an economy. To determine the size of area requisite in this case for producing a determinate quantity of bark, we have no data—as there is no data—as to how much bark is collected per desatin. For crushing the bark many fabriques have steam crushing machines.

'With regard to the value of other forest materials requisite for the leather fabrication, we must remark that in the centres of this fabrication, Bolchov and Jeletz, the prices of birch and oak wood vary from 8 to 10 roubles; aspen wood, from 6 to 7 roubles; tar, from $1\frac{1}{2}$ to 2 roubles; pitch, from 1 to $1\frac{1}{2}$ roubles. The prices for the last two articles will soon be lower, because the considerable use made of these for the wheels of carts for transporting goods will soon be less, as with the building of railways the number of carts going any great distance will not be so great.

' The substitution of antracite for wood is being introduced very slowly in Jeletz, as they say it costs 40 kopecs per pood. At this price it of course cannot replace wood, which even in the Jeletz district costs about one-fourth of that less. With the opening of the Jeletz-Briansk railway antracite will certainly be cheaper, and forest materials will get dearer, so that the change must take place; but there is no danger to the sale of forest materials from such a substitution, because the demand for them is increased.

' With regard to the substitution of turf for wood, although this is possible in some districts, they seldom attempt the working of turf. The greater number of the peasantry consider it a sin to heat with earth; and if with the increase of workmen after the railroads are completed they should work it in large quantities, meanwhile much of the turf riches will be washed away by water, and will be burnt out by the carelessness of the local population.

' By this means the competition of turf and coal,

however desirable, is almost impossible; unfortunately
other kinds of fuel: kiziah (dried cow dung), louzga, and
straw, can be much used; this will lead to the exhaustion
of the fields, which will be more serious than the exhaustion of the woods; for however fertile the land may be, with
such economy it may be soon converted into unproductive
ground.   If the products of the first necessity are getting
every year dearer, the principal reason of this dearness is
not so much financial difficulties as the wasteful
system of carrying on agricultural economy.

Sales from the Crown Estates of the Orel Government.

'The wood trade in the crown estates, or as they say, sales
from the crown estates of the Orel government, present
facts worthy of attention, and partly confirm our views
of the importance in the wood trade of the wood belt
of the Orel government—

In 1863 were sold 466 desatins for 16,785 roubles, or 36 roubles per des.
   1864, no information of the
         number of desatins
         sold,   -   -   -   23,251   ,,
   1865, were sold 945 desatins for 34,524   ,,   35   ,,
   1866, ,,   692   ,,   38,625   ,,   55   ,,

'That is, the prices in the course of four years were
nearly doubled.   The principal mass of the wood sales is
in the wood belt of the Orel government, where there is
rather an excess then a want of forest materials; and
therefore the whole area of clearings offered for sale
was not bought.   For instance, in 1863 940 desatins
were offered for sale, but only 466 desatins were bought;
in 1865, 2004 desatins were offered, and 945 were bought;
in 1866, 2094 desatins were offered, and 692 were bought.
'Comparing 1865 with 1863, we see that double the
area was then bought; but the price per desatin remained
almost the same, consequently the increased demand did
not lead to higher prices, because the quantity offered for
sale exceeded the demand.   With a more commercial

system of sales, of which I have already spoken, the prices per desatin would have risen much more rapidly. The increase of the income in 1865 was principally from the increased demand for forest materials for constructing railways; with the termination of the construction this demand will somewhat diminish; and therefore for the next three years one must expect for the government of Orel a less progressive demand than in the last three years. In the government of Orel, besides the sales not accounted for, sales are made with payment for the manufactured article, and likewise sales are made in advance per contract, with increased payment after the expiration of a few years. In view of the rapid increase of prices, and greater demand for crown woods, to which we have frequently drawn attention, the sales of wood in advance for long periods cannot be considered advantageous; but with regard to sales the material to be paid for when manufactured, it appears to us that the simple sales, the advantages of which have been proved by the experience of the last years, deserve a preference, particularly in such places as the woody parts of the government of Orel, where the rapid increase of demand and sale is guaranteed.

'The above ciphers of the income are taken from the general amount of income in order to show more accurately the change of price per desatin of forest area generally, but the total income of 1863, 1864, and 1865 were as follows :—

| | 1863. | 1864. | 1865. |
	Roubles.	Roubles.	Roubles.
Sold, - - - -	1,9561	25,817	48,373
Sold by lower estimate, -	5,326	6,341	15,955
Without money, - -	245	619	335
Received rents from estates,	5,656	5,602	6,480
,, arrears, - -	578	1,475	682
Duty for billets, - -	139	143	168
Interest, - - -	378	570	944
	31,883	41,167	72937

' With the general extent of forest estates under crown management of 344,832 desatins, there was received of income—

In 1863,	-	-	-	-	9 kopecs per desatin.
1864,	-	-	-	-	12 ,,
1865,	-	-	-	-	21 ,,

But taking into account only the area covered by forests, viz., about 255,533 desatins, we find the amount received—

In 1863, 24,649 roubles from the forest area, or 10 kopecs per desatin.
1864, 33,491 ,,   ,,   13   ,,
1865, 65,755 ,,   ,,   21   ,,

In this case we take into account the wood sold, and also that sold at a reduced price, and free duty for billets, and interest for money exacted in selling the timber.

' Of the estates of the government of Orel deserving notice from the value of the yearly clearings, the first one is not far from Orel, the crown estate Poslovskaio, of the extent of 236 desatins 1740 fathoms. In this estate there are 221 desatins of forest area, and 15 desatins of waste land. Of the forest land 113 desatins are under pine trees ; the remainder is under oak growth. With regard to age 40 per cent. are mature, 40 per cent. are young plantations, and 20 per cent. are of medium growth. There is one desatin sold yearly from this estate, at the price of 400 to 500 roubles. This valuable estate remains until now not organised, and it is guarded by foresters. There is no word yet about artificial renewing of the estate, notwithstanding its great revenue. The income from this estate is appropriated to the Alexandrina Female Institution.

' The greatest number of estates in the government of Orel, giving an annual income from the sale of yearly fellings, are concentrated in the first Briansk forest district. Of the general area of the forest estates of this forest district of 61,399 desatins 2,230 fathoms, the estates having a full sale cover 59,365 desatins 1,012 fathoms. In this area are included the crown estates:—Souponev, 7,927 desatins 660 fathoms; Soensk, 7,485 desatins 2,251 fathoms; Koulnev, 16,819 desatins 2,364 fathoms; Koritchijko-Krilov 3,576 desatins 2,209 fathoms; Vorpomorsky, 13,513 desatins 1355 fathoms. In these estates there is of forest 41,249 desatins, of which there are under pine, 18,414 desatins; under fir, 7,616 desatins; mixture of coniferous and broad-leaved trees, 8,386 desatins; under broad leaf, 5,167 desatins. In each of these estates some parts of them lie at 5 and 10 versts from the most convenient roads for sales; other parts of the same estates lie still further, so that the division of the mass of wood by estates, as it exists at present, is not satisfactory either with regard to forest trade or economy. With regard to trade in wood, since into one whole are joined parts differing between themselves as to expense of cartage of timber, this is inconvenient in making estimates in an economical point of view,—because being in contact with each other these estates require their boundaries to be cleared, whereas if the whole area of these estates were taken as one economical unit then all the trees of the same section by one numeration could be divided according to convenience of cartage; and then the boundary lines would be changed for division lines, which would give great convenience in making the estimates, and in an economical point of view would require less labour in clearing the cuttings.

'Above we have remarked that the annual sales are preferable to the contract sales for cuttings for a series of years, and we adduce in confirmation of this the following:—In the first Briansk forest district is situated the Polpinsky crown estate, which is on lease; its total area is 4,652 desatins 760 fathoms; of this area their are under wood 4,016 desatins; annual cuttings, 58 desatins; giving a gross revenue of 676 roubles, 46 kopecs, or 11 roubles per desatin; whereas the average value of a desatin of clearings in the government of Orel, as we saw

above, was from 30 to 50 roubles; moreover, this average
price is rapidly increasing, and the lease rent remains for
a number of years the same.   In this forest district there
is likewise the estate of Koritchijko-Uralov (in the
management of the Briansk forest officials) of 2,850 desatins 1,835 fathoms.   Besides the crown estates there are
peasant estates, the total area of which is 4,474 desatins
396 fathoms.

'The first Briansk forest district is the principal one in
the government of Orel, as well from its present condition, as
from the future which it will have with the opening of the
Smolensk-Orel railway, which runs through it for several
versts.   From the value of the cuttings in the above-mentioned estates of this forest district Koulkovsky is remarkable.   The average price per desatin on this estate is 87
roubles, but frequently their are parts where the price
per desatin is above 100 roubles.   One meets likewise on
the Soensk estate with desatins much above the average
price per desatin in the government.   In this estate some
desatins attain sometimes to 140 roubles.   The Soensk
and Koulkovsky estates are also known under the names
of first and second parts of Polninsky estate.

'The Karatchev forest district has almost like importance with the first Briansk, because from most of the
estates of this forest district there is a possibility of transporting forest materials by railway (not over 80 versts),
with a cartage of 5 to 20 versts.   Besides the district town,
Karatchev is a local consumer, and is situated at a
distance from the principal mass of forests of about 20
versts.

'The principal estates of this forestry are :—Recetitzk,
5,020 desatins 1,380 fathoms ; of this there are under forest,
4,645 desatins 2,355 fathoms; on lease, 118 desatins
1,580 fathoms ; in the management of the forest guards,
97 desatins 400 fathoms ; under the remaining appurtances, 5 desatins 2,880 fathoms ; waste area, 152 desatins 1,985 fathoms ; under pine, 685 desatins ; under
fir, 45 desatins ; under mixed kinds, 2,657 desatins ;

under birch and alder, 1,226 desatins. At the annual cuttings on this estate one meets with fellings, the value of which often attain 100 roubles.

'The bordering estate, Poldevsky, 16,263 desatins 1,875 fathoms, is under similar conditions of sale. The division of these estates from each other has no importance whatever in an economical point; both the estates are organised; the bordering corner lines between them are quite useless. It would have been much more useful in a commercial view for making estimates if both had been grouped together with regard to the distance of their quarters or sections from the town of Karatchev, and from the line of railway, &c. Such grouping of quarters of this mass of forest, in an economical point of view, is much more practical than the division of the estates according to the boundaries of the general survey. The whole extent of the Karatchev forest district is 25,142 desatins 2,194 fathoms, of which there is in the possession of the peasantry only 1,525 desatins 2,384 fathoms.

' The first Briansk and Karatchev forest districts occupy in the government of Orel the first place among all the forest districts of this government with regard to the facility of sales of forest materials; and likewise the convenience of living for the foresters:—Land, houses, nearness to railways, and the three towns Briansk, Karatchev, and Orel, are of course not unimportant privileges. For these two forest districts it is of the greatest importance to have roads for carting timber or cuttings, which should by the shortest way connect the clearings with the railways, and would afford the cheapest means of conveying the timber from the most distant parts of these forestries.

' The third Briansk forest district is remarkable from having the Okonlitsk estate, upon which many look as an immense reserve of forest riches which may readily rot in the wood without any profit to the crown, and which require, therefore, an early and speedy sale. We do not adopt this view.

M

'This estate occupies 53,853 desatins, and must be divided into two principal parts: a south-western near the river Iput, and a north-eastern near the railway from Smolensk to Orel. By this railway the distance to Briansk is not above 40 versts from the north-eastern part of the Okonlitz estate. A considerable development of the wood trade in Briansk, and the constant decrease of private estates, will in a short time oblige the wood dealers to have recourse to the Okonlitz estate, the more so, that with the opening of the railway to the western Dwina, they can get by it goods for Riga. The products of forest technical manufacture can be most profitably obtained from the Okonlitz station, because, being near a part of the country where there is no wood, there is no local demand for forest materials, and therefore everything that is unprofitable to transport by rail may be manufactured; and with the abundance of pine plantations, of which there are 1,600 desatins on this estate, the abundance of pitch to be obtained from fallen trees allows a possibility for the pitch business to exist for several years coming. Towards the end of this century there will ensue a rapid increase of income from the Okonlitz estates, if until then the sales be made on the general principle of fellings for immediate payments, without a hasty seeking of buyers on however exclusive conditions. This is the most desirable system of sale from the Okonlitz estate, but with the payment of duty, not from the quantity of materials manufactured, but from the number of fallen trees, for only in this case will the felled timber be most economically converted into goods.

'In the first Tronbteheosk forest district is worthy of remark from its size, the crown estate Ouspenskaia, of 30,548 desatins. This estate is remarkable in that it has on lease 1,561 desatins, and waste area, 5,422 desatins; of forest, 23,319 desatins; of which pine occupies 12,061 desatins, and the remainder is under different broad-leaved species. The total crown estates in this forestry is 33,199 desatins, and there are 10,046 desatins belonging to peasants, so that the whole amounts to 43,246 desatins, 145 fathoms.

'The third Tronbteheosk forest district consists of estates in the possession of the peasantry, the whole area of which is 6,356 desatins 677 fathoms.

'In the second Tronbteheosk forest district the principal forest mass is the Novonikolskaia crown estate, measuring 2,805 desatins 1,400 fathoms. In this estate the forest area is 25,106 desatins, of which 9,565 desatins are under pine, and 9,142 desatins are occupied by a mixture of coniferous with leafy trees; 3,539 desatins are under birch, and 2,172 desatins are under alder. An extensive space in this estate is under waste, 1,495 desatins 1,798 fathoms; and under lease there are 826 desatins 1,470 fathoms.

'Into part of Seosk forest district enter first the forests of the Seosk district (which we consider to belong to the Dwina river forest belt) amounting to 10,710 desatins 1,294 fathoms; and secondly, the forests of the districts Kromin, in which there are forests under crown administration, 1,395 desatins; and Dmitrovsk, in which are forests under crown administration, 770 desatins 833 fathoms in extent.

'The remaining three forest districts—Liven, Jeletz, and Orel, are composed almost exclusively of peasant estates—Liven, forest district of estates of the Liven; and Malo-Archangelsk, districts of 7,645 desatins 696 fathoms; Eletz, forest district of peasant estates of the districts of Orel, Bolchoff, and Mtzensk; and there is one crown estate in the Orel district.

'The general distribution by districts of the crown and private forests is—

		Crown.	Private.
3 Briansk forest districts	in Briansk,	161,400 des.	234,897 des.
3 Tronbteheosk, ,,	Tronbteheosk,	87,890 ,,	214,525 ,,
Karatcheff, ,,	Karatcheff,	25,142 ,,	36,566 ,,
Seosk, ,,	Seosk,	10,710 ,,	51,125 ,,
	Kromin,	1,995 ,,	7,840 ,,

Seosk forest districts—		Crown.	Private.
	Dmitrovsk,	770 des.	37,429 des.
Liven, ,,	Liven,	7,645 ,,	10,048 ,,
	Malo-Archangelsk,	1,945 ,,	1,895 ,,
Jeletz, ,,	Jeletz,	15,703 ,,	34,471 ,,

'Such a preponderance of private forests over crown in the government generally, and in some districts in particular, places rural economy in great dependence on private forests ; and in the meantime the conservation of private forests is the weakest point of forest economy. In the present year, by the direction of the government Land Administration, a pamphlet was printed for the assembly of the Land Administration, in which it is stated that the destruction of forests may lead in one case to the decadence of the leather trade in consequence of the want of bark and firewood, and on the other hand to the exhaustion of the cultivated lands, in consequence of the constantly increasing consumption of dung and straw as fuel. In this pamphlet it is stated that the decadence of the private wood trade proceeds from the general destruction of private forests, and from the sale in the crown estates in small parts of annual clearings, in consequence of which the bulk of the purchasers have turned to crown forests.

'If we take into account the revenue from private forests, and the number of wood dealers preparing in them forest materials, we may suppose that the wood trade, and the income from wood, is falling off; but if of the forest trade and the income from forests we judge from the amount of income from the crown forests then we can maintain that the forest trade is not falling, and that the income from the forests is constantly increasing—because it is only under such conditions the forest income in the government of Orel could in the course of three years double itself as it has done. This is sufficient to show the characteristic difference between private and crown forests of the Orel government.

'In the above-mentioned pamphlet an idea is propounded

as to the necessity for the peasantry to occupy themselves with the planting of forests; one cannot but wish for this idea a realisation in practice, particularly if with the trouble taken about the future forests the existing forests be not forgotten.

Remarks on the Wood Trade of the Government of Kursk.

'The forests of the Kursk government are generally not taken into account in speaking of the Dnieper basin; the reason is that the Kursk forests are situated too far to the east of the general mass of the Dnieper forests. The greater part of the Kursk forests are on the confluents of the river Seyma, which forms a part of the Dnieper basin; and cutting in, as it were, in a long and narrow stripe between the Oka and Don basin, it is situated at the northern, and at the same time, the most eastern part of the woodless belt of the Dnieper basin.

'From the banks of the river Seyma commences that part of the Dnieper basin country where there is little wood; likewise from the river Seyma commences in a southern direction the preponderance of oak, and that of pine ceases; finally, from the banks of the Seyma commences that part of the Dnieper basin in which there is not sufficient wood for local consumption. Further east than the government of Kursk forest materials are not taken from the Dnieper forests; and only south to the governments of Kharkov, and partly Starropol and the land of the Don-Cossacks, are taken forest manufactures of oak; and with regard to building materials, boards, and planks, these are taken from Glonchov and Novgorodseversk, districts in the government of Tchernigov, through the government of Kursk (Pontivle district), to the districts of Soumnic and Lebedian in the goverment of Kharkov. In this way forest materials very seldom get to the towns of Achtiska and Bogodonihov, and still more rarely to Kharkov, so that the boundary of their spreading south-east may be taken the post road from Kursk to Kharkov and further to Constantinograd. In Constantinograd commences the supply of forest materials by land which were floated down the Dnieper to the ports of Krementeboug and Ekatherinoslav.

'The forests of the government of Kursk have their principal markets within the boundaries of the government; and in consequence of its having little wood the sales are very profitable. Generally, for the Kursk government forest materials are brought from the government of Tchernigov for the districts of Pontivle, Ruilsk, Soudjan; and for the districts of Kursk, Lgov, and Fatej, from the districts of Tronbteheosk and Seosk of the government of Orel. The cartage from the government of Orel is principally done by the peasantry of the Dmitrov district, whose position at about the half-way gives them the means of using their own provender, and of not requiring to be long absent from their homes. For both of the cartages (from the governments of Orel and Tchernigov) there is a different future. The cartage of forest materials from Tchernigov government to the districts of Pontivle and Ruilsk in the Kursk government, and to Soumnia and Lebedran, in the Kharkov government, must diminish, because the Kursk-Kiev railway, passing through the Tchernigov government, facilitates the transportation to these districts of a great part of the forest materials by railway. But the cartage of forest materials from the districts of Tronbteheosk and Seosk in the Orel governments to Fatej and other districts of the Kursk government, will increase, because the Kursk-Orel railway will not only add to the demand for forest materials in these districts, but it will give no facilities for transporting them cheaper, because the cartage is by a much shorter road than by rail.

'Going to the Tronbteheosk and Seosk districts for forest materials, the peasantry take some goods from the Kursk government, principally grain, or for small dealers, different spices, fruits, and wine, and on their return bring principally boards, and small binding materials.

Under favourable circumstances they manage to make six turns (a turn is considered the sale of the wood.)

'Favourable circumstances consist in fine weather and the quick purchase of forest materials in the forest belt, and in the quick sale of these in the parts requiring wood.

' With regard to the fine weather they are guided by the traditions of the elders, and as the medium length of each half way is about a week, therefore for such a short time the predictions of the local meterologists are generally not far wrong.

'For expediting the purchase of forest materials in the forest belt they apply to the some dealers, or joining together, buy a part of a wood, divide it among themselves, tree by tree, and then each fells his own and carts it away with his own horses. As guard and manager on the spot there remains generally the most respectable member of this company, and he receives for his trouble a remuneration in forest materials; remunerations in money are considered insulting. The principal condition for the success of this trade are rapid sales in places requiring wood. For this object they find purchasers beforehand, and make an agreement as to what materials, and for what price, they are to bring. Frequently the materials are brought to market, and if there be no buyer at a remunerative price (which sometimes happens when there is great supply), they have to feed their horses several days in the expectation of a buyer. But this is unprofitable, because buyers appear generally only on market days, of which there is only one or two a week, and consequently the provender of horses would cause great expense; and they sell their trees to a richer dealer who can wait for a purchaser. Such a sale does not bring the same profit as the sale to a consumer would. With such a supply trade the supply may not be proportionate to the demand, and in such cases the prices fall to the last extreme. In such cases one may see the very same kind of materials sold by different dealers on the same market days asking prices varying from 1 to 2 or 1 to 3. This proceeds from the fact that one has been in the market a long time and not sold his goods, and lowered the price; while another has only just come, and is not convinced of the excess of supply, and therefore keeps to the normal prices; but several days will pass, and he will lower his prices more than his neighbours.

'The want of capital for buying forest materials in such cases has created something like commission yards. For a certain allowance of part of the materials, or of the realised sum for their sale, they put up the unsold materials with an acquaintance; he is empowered to sell them; and they call for the proceeds at their leisure. Many capitalists, wood dealers, have commenced by such commissionship in the wood trade in the governments of Orel and Kursk.

'With regard to the importance of the crown forests of Kursk government in the wood trade we remark that the situation of these forests is very favourable as regards sales—1st, Because crown forests are in better condition than private forests, although I must add that many of the latter are very satisfactorily preserved; 2d, because the crown forests form almost half the forests of the Kursk government; and 3d, because the greatest part of the crown forests are in districts that have less forests, and are more distant from the woods of the Dwina river belt governments of Tchernigov and Orel. We therefore suppose that cases of unsuccessful sales of wood cannot be explained by the competition of private forests, as many allege, but that rather the contrary takes place, because sales to peasants at a reduced price or one much lower than the market prices made to a considerable extent, must really tend to reduce prices, or stop for a time the sale not only from private, but even from crown estates, because these reductions draw after them a decrease of demand in the markets, and often consequently limit the extent of the wood trade.

'All the forest estates under crown management in the Kursk government form five forest districts :—

	BELGOROD.		KARATCHAN.		RUILSK.	
	Desatins.	Faths.	Desatins.	Faths.	Desatins.	Faths.
Forest area, - -	27,220	1,492	65,836	40	15,582	—
Lease sections, - -	—	—	1,282	2,350	—	—
Land belonging to guards,	126	1,070	84	56	37	1,113
Under appurtenances,	584	100	43	1,000	459	205
Waste land, - -	261	990	178	1,373	138	2,180

	LGOV.		KURSK.		TOTAL.	
	Desatins.	Faths.	Desatins.	Faths.	Desatins.	Faths.
Forest area, - -	22,736	699	38,328	1,983	169,703	1,178
Lease sections, - -	364	20	—	—	1,646	2,370
Land belonging to guards,	32	20	40	50	319	2,289
Under appurtenances,	982	667	1,287	758	3,359	830
Waste land, - -	454	752	501	718	1,532	13

'The general area of private estates is considered to be 200,000 desatins. In this are pear plantations, but the system of using these does not give sufficient reason to include them in the forest area; and besides this there are waste spaces, so that of actual forest area in private estates there is not more than in the crown estates, that is about 170,000 ; and the sum total of forests in the government is about 340,000 desatins, which for 1,827,000 inhabitants, gives a very small quantity. The greatest number of forest estates in the Kursk government are appropriated to supplying wood to the peasantry at a low price ; the sales from the crown estates are only on 11,222 desatins 965 fathoms. From the sales of forest materials from this space was received—

In 1864,	-	-	-	-	-	14,141 roubles.
1865,	-	-	-	-	-	19,740 ,,
1866,	-	-	-	-	-	21,066 ,,
1867, (before the end of the year),					20,236 ,,

' Consequently in the latter years the gross income per desatin, exclusive of waste land and appurtenances, was about 2 roubles. But if we consider the income from sales in 1861 realising only 6,936 roubles, we remark that in

six years the income nearly trebled itself. In 1867 there
were fellings for ready money sales on some estates in the
Kursk government—and this deserves the attention of
specialists, because it presents data not void of some
interest in discussing the subject of estimates.

'In the crown estates of the government of Kursk
estimates were made, and in some estates there was not a
full sale, the reason assigned was the high estimate of
fellings; and therefore to the contracts of 1861 the
clearings were offered by a new estimate.

'The estimates and the market prices present the
following figures :—

		First Valuation.	Second Valuation.	Market Prices.
	Desatins.	Roubles.	Roubles.	Roubles.
Belgorod forestry and district—				
1st Staritzk estate, - -	36	2,299	2,299	3,400
Graivoron district—				
Kovensky, - - -	13	1,147	1,000	1,634
Shiskoffsky, - - -	2	122	139	260
Korotchan forestry, Novoshelky district—				
Michailov, - - -	15 ·	2,317	2,124	3,801
Dmitrov district—				
Kostin estate, - - -	14	478	300	526
Ruilsk district and forestry—				
Alexayeff estate, - -	2	140	123	140
Pontivle district—				
Botchagan estate, - -	1	214	203	351
	83	6,717	6,188	10,112
Belgorod forestry and district—				
2nd Staritzk estate, -	31	6,839	2,634	4,025
Lgoff forestry and district—				
Kopesheosk, - - -	30	2,485	1,278	1,285
Borisosk, - - - -	2	540	320	329
Dmitrov district*—				
Korobkinsk, - - -	22	6,815	2,700	3,010

* Of this district is not shown the Popoviusk estate, from which there was sold for
711 roubles, which, with 19,524 roubles, will make 20,255 roubles, shown above in the
income for 1867.

		First Valuation.	Second Valuation.	Market Prices.
		Desatins.	Roubles.	Roubles.	Roubles.
Dmitrov district—				
Menshikov,	· · ·	15	340	180	220
Olchov,	· · · ·	4	211	110	111
Pogodin,	· · · ·	7	389	260	280
Kursk forestry, Oboyan district—				
Medvensky,	· · ·	1	155	100	152
Total,	112	17,233	7,582	9,412
Grand total,	24,073	13,770	19,524

'Looking over these ciphers we see that for the 83 desatins the first valuation is not only not higher than the market prices, but even much lower than this; the second valuation is still lower; but with regard to the 112 desatins the first valuation is nearly twice as high as the market price, and the second valuation is lower than the market price by 24 per cent. Separating each estate, the difference presents great fluctuations, but in general the first valuation approaches nearer to the market price valuation, the difference between them is 4,500 roubles, whereas the difference between the market price value and the second valuation is 5,800 roubles.

' If one were to judge by the per centage addition at the sales, then the second valuation will of course give other results than the first. Many think that the per centage addition at the sales may be considered as a sign of the great activity of the local management, but we think that per centage additions may be rather considered a sign of the incorrectness of the estimates, or that the estimates were incorrectly applied; the removal of the unfavourable results from both the incorrectnesses, we attribute to the circumstance that the rules confirmed by the Emperor for ready money sales of forest materials by public sales are quite practicable, and are more suitable to the present state of our forest economy,

*Remarks on the Wood Trade of the Government of
Ekatherinoslav.*

The government of Ekatherinoslav is divided into two
principal parts—1st, the Don basin, and 2d, the Dnieper
basin. The local wood markets in the Don basin receive
wood from the banks of the Volga and Don; and those
in the Dnieper basin, principally from the ports of the
Dnieper. In the Don part are situated much more
important wood marts than in the Dnieper, as much
judged by the higher price as by the variety of forest
materials required in the markets.

' The wood trade in the Dnieper part is centred principally at the ports on the Dnieper above the rapids, particularly at Ekatherinoslav. The reason is—1st, because
the port of Ekatherinoslav is the nearest to the great
eastern part of the government ; 2d, because the purchase
of forest materials before the rapids is much more profitable, as high prices have to be paid to traders for the risk
of floating over the rapids. Therefore from Ekatherinoslav, as from ports lying before the rapids, forest materials
are taken in every direction—to the government of
Poltava as far as Constantingrad, to the Taurid government as far as Bezodiansk, and the Kherson government as
far as the river Ingul. All this transportation is done by
land, and principally during summer. In winter the
roads are not reliable. Although there are are snowy
winters, more frequently the snow appears and disappears
several times during winter, and on sledge roads one
cannot count even on January. One strong wind from
the south or south-east produces a thaw sufficient for
melting away the snow. Exposed places and strong
winds cause the snow to be distributed unequally. In
the *steppe*, in exposed places, snow barely covers the
ground, but in ravines and thickets it forms a layer of
considerable thickness. Therefore one cannot rely on
winter roads for the transport of forest materials, so the
winter is here considered the most unfavourable for

the transport of forest materials. Successive thaws
in winter, and rains in spring and late autumn, make
the transport of forest materials more difficult on
wheels. The black soil gets wet through for some
eighteen inches, and the cartage of heavy loads is
difficult. Only in summer, the end of spring, and
beginning of autumn, áre the roads passable for burdens ;
the dried soil becomes even and hard, and at this time
loads can go with the same ease as on a macadamised
road. Rains during the summer produce mud about
three inches deep, but this dries soon ; one or two sunny
days, and a strong wind, and the road becomes practicable
—this is the time most suitable for transporting burdens,
and consequently forest materials. But at this time field
labours take place, and the want of means make the
transport of forest materials unprofitable ; the end of
summer and the beginning of autumn, set free from field
labours a considerable number of transport means, but at
this time commences the cartage of the products of rural
economy to the ports on the Dnieper, and only after the
sale of these products are the transport means free.   In
order not to return home empty from the ports of the
Dnieper forest materials are bought and taken back to
Ekatherinoslav, and the neighbouring governments.

  ' By such means the forests of the government of
Ekatherinoslav meet with difficulties in transportation, so
that they sometimes remain partially unsold, notwithstanding the general want of forest materials.

  ' In the government of Ekatherinoslav the competition
of different substitutes for wood has attained greater
development than in other governments : stone buildings,
huts built of clay and kizik (cow's dung), lougia, reeds, dry
grass (called burian), straw, coal, and anthracite, and
finally with a long and hot summer, and winters not cold,
these are competitors with forest materials in this government. But notwithstanding this competition the prices
of forest materials are constantly rising. Although in the
eastern part of the Ekatherinoslav government, and the

parts joining it of the land of the Don Cossacks, there are, as is generally known, beds of coal and anthracite, and the working of the last has received in these places greater development than in other governments; still, the prices of wood materials here attain their highest delelopment. In Taganrok, Rostov, and Nachitchevan, lying not far from the centre of the coal district of Russia, the price of firewood varies between 25 and 36 roubles, and the rise of price in Taganrok to 36 roubles took place in the last year, notwithstanding that in the neighbourhood the coal industry was developing itself.

'The crown forests of the government of Ekatherinoslav are divided into the following forest districts :—

	Forest area.	On lease.	Ground of foresters	Appurten- ances.	Waste land.
	Desatins.	Desatins.	Desatins.	Desatins.	Desatins.
(Basin of Dnieper.)					
Novornoskovsk,	6,007	278	262	1,472	1,740
Ekatherinoslav,	4,364	3,717	—.	—	226
Verchnedreoposk,	4,924	—	180	432	2,918
Parlograv,	5,373	130	108	—	494
(Basin of Don.)					
Slavianoserbsk,	4,364	—	288	—	226
Rostov,	1,064	—	47	—	—
Bachmut, -	2,688	—	110	—	61
Velikoanadolsky,	1,746	2000	60	447	39
	30,530	6,125	1,055	2,351	5,704

'In the government of Ekatherinoslav there are in all 101,147 desatins of forest, but in this amount the crown forests are entered entire—that is, with their lease grounds, lands of foresters, &c. The same must be said of private estates, in which are included the commons on which trees grow.

' Many of the crown estates, by their small size—a few square fathoms—have no importance in the wood trade, and are incapable of repaying the expenses of watching. The same must be said of the forest plantations on which up to this time trees have been planted. The following is the revenue from the great estates—

	1863. Roubles.	1864. Roubles.	1865. Roubles.	1866. Roubles.
Sold for, - - - - -	5,055	5,264	4,384	4,956
Disposed of at a reduced tax, -	10,049	14,519	14,619	15,225
Given gratis, - - - -	1,724	1,475	2,544	—
From leases and sundry, - -	3,321	1,971	5,151	4,987
Received arrears, - - -	11	111	56	515
,, duty for billets, -	50	50	48	47
,, interest, - - -	55	103	86	80
	17,854	22,999	25,700	25,136

' The smallness of the revenue from the sale of wood is principally because the greater part of the wood is destined to be given to the peasants at a reduced price, in consequence of which there is not a proper state of affairs in the wood trade on the local markets. As a characteristic of the difference between the sales at a reduced price and free sales, I adduce the following ciphers :—

' From the sales at a reduced price or tax in 1866 was received 20,181 roubles, which, with the total area of forests of 48,496 desatins, makes for each desatin 41 kopecs income ; but from sales much more was received :

' For instance, in the Rostov forest district there is a crown estate, Leontiev-Boerak, of which the area is 1,064 desatins, and the revenue received from it was—

From sales, - - -	3,179 roubles,	52 kopecs.
2 per cent. money, - -	64 ,,	65 ,,
Stamp duty, - - -	21 ,,	—
Fines, - - - -	86 ,,	65 ,,
For illegal felling, - -	3 ,,	20 ,,
	3,354 roubles,	43¼ kopecs.

That is, more than 3 roubles per desatin, or eight times more than where the timber is given at a reduced rate.

Remarks on the Wood Trade of Kherson Government.

'The government of Kherson belongs to those few governments which receive the wood required for local purposes from beyond the frontier; but in the government of Kherson foreign wood is used only in small quantity, while the Dnieper private forests are not quite exhausted. Forest materials are brought into the government of Kherson by three ways — by the Dnieper to the eastern part of the government, by the Dwina to the west, and principally to Odessa from the government of Podolia and from Galicia, by land from the governments of Volhinia and Kiev to the north-western part of the government, to Odessa and Nicolayev by water from Kherson and from the mouth of the Dwina.

'The principal demand is for timber of large dimensions fit for sawing and shipbuilding. The development of our foreign trade, increasing navigation, and ship-building, increases rapidly the prices paid for the larger dimensions; and after this, though somewhat slowly, the prices of other sorts rise, and in the meantime the possibility of satisfying the want of forest materials, particularly of large dimensions, is diminishing from year to year as the forests get exhausted. The renovation of the Black Sea fleet will very likely take place soon, for one cannot suppose that this fact will belong to the distant future. With every rumour of this event the prices for large dimensions rise even now, when these rumours have no foundation, but when they become real then the rise of prices will be immense. Some wood dealers, buying estates with large timber, put off the working of them waiting for this event as likely to occur at least within the next ten years.

'The Dnieper forests present the chief source of the most convenient and profitable supply of forest materials for our shipbuilding in the Black Sea. If such a calculation be not considered unprofitable for wood traders, then the more so in imperial forest economy must one reckon on the possibility of a change in the demand for forest materials, depending on political events which have taken place, or which may take place in the future. Even if we do not calculate on this temporary rise in the prices, in any case the forest economy in the Dnieper forests being directed principally to the attaining timber of large dimensions, promises very profitable results, because the sales from private estates of firewood and small timber is increasing every year. By increasing the mass of felled timber private individuals try to cover the deficiency of income arising from the exhaustion of timber of large dimensions. The sales of timber of large dimensions from private estates is visibly diminishing, and with regard to them crown forests meet every year less and less competition.

' Forest materials floated down the Dnieper to Kherson are taken by sea to the ports of the Crimean peninsula, almost to Theodosia, principally timber for sawing, and fit for shipbuilding. To Nicolayev wood of all sorts goes, and from thence higher up the Bug to Vojnesensk, principally boards and building timber, and Odessa is supplied with wood of all sorts. As a characteristic of the dearness of forest materials in Odessa may be mentioned the sale of firewood, which is sold in such small quantities, that in the wood markets, kiziah (dried dung), reeds, and other fuel, is sold. A few billets are sold for 5, 10, and 15 kopecs to the poorer classes, to serve as chips to light the fire. In such a sale a fathom of wood produces not less than 35 roubles, and such a price exists in a town where the consumption of coal is much developed.

' To Kherson, from the upper parts of the Dnieper, wood is floated down only by wood dealers, and here it is purchased wholesale by other dealers for sale in the Kherson, and in the ports of the Dnieper, or for shipping by sea to the above-named places.

' The principal demand is for timber for sawing, and as workmen are dear, two sawmills have been erected at Kherson. Their importance is not the same here as in the

N

forest country : there they facilitate the sale of forest
materials, whereas here they only facilitate the cutting of
these, which even without this have a ready sale ; therefore
the sawmills of the government of Kherson do not form a
necessity of the wood trade, but are hired to saw up the
logs, as workmen are hired generally, and this limits the
importance of sawmills in the wood trade.   With regard
to the influence of cartage on the wood trade I must say
that in this respect the government of Kherson is very
similar to the government of Ekatherinoslav.

'The railway in the Kherson government passes principally near those parts where the crown forests are grouped,
and where the foresters reside, viz. :—Tiraspol, Ananiev,
Novomirgorod, and Krilov. In consequence of this the
forests of the above-named forest districts are of great
importance, and the forest districts themselves, from the
convenience of life for the foresters, deserve the particular
consideration of those that wish to become foresters.

'The government of Kherson is divided into the five
following forest districts :—

	Forest area. Desatins.	On lease. Desatins.	To foresters. Desatins.	Under appurts. Desatins.	Waste. Desatins.
Kherson, - - -	1,028	1,533	37	410	158
Odessa-Tiraspol, - -	4,698	15,321	113	3,267	758
Ananiev, - - -	6,914	355	216	501	79
Novomirgorod, - -	12,565	—	428	1,055	330
Krilov, - - -	9,353	—	266	415	191
	34,768	15,209	1,054	5,648	1,516

With regard to revenue there are the following data :—

	1863. Roubles.	1864. Roubles.	1865. Roubles.	1866. Roubles.
Sold, - - -	4,439	9,130	5,406	44,268*
Sold at a reduced price,	1,196	291	576	252

* The forests of the military colonists were added in 1866. The income from the
forests of the military colonists amounts to about 35,000 roubles.

	Roubles.	Roubles.	Roubles.	Roubles.
Given gratis, · ·	57	60	61	2,052
Loans, · · ·	4,898	6,402	5,730	11,735
Arrears, · · ·	1,419	432	291	—
Duty for billets, ·	61	53	26	—
Interest, · · ·	89	105	69	53
Reeds, hay, &c., sold,	2,174	3,362	3,882	1,262

'In the Novomirgorod forestry, district of Alexandria, is situated the estate of Tchernoless, or black forest, remarkable in New Russia for its great size (3,561 desatins 227 fathoms), for the government of Kherson, as well as the age and denseness of the plantation, and besides this, by the different archæological tronvailles and researches near this estate. The trading importance of this estate is characterised by the sale of 98 desatins for 27,794 roubles. In this forestry and district is the Neronbaiev estate of 2,827 desatins 220 fathoms, from which were sold 2 desatins for 1,120 roubles. In consequence of the importance of these two estates we consider it necessary to relegate some particulars about them to a separate article, as likewise others relating to the Tchouticansby estate, Krilov forest district, from which 66 desatins were sold for 18,952 roubles.

'By the value of the cuttings are distinguished likewise the following estates :—

'1. Boboskhov, 173 desatins 2,280 fathoms. Of this general area there are under forest, 140 desatins 1700 fathoms ; under plantations by forest guards, 5 desatins 400 fathoms ; under appurtenances, 26 desatins 1,500 fathoms ; waste, 1 desatin 1,080 fathoms. Of the forest area there are under oak, 80 desatins ; yoke-elm, 20 desatins ; different leaved trees, 32 desatins ; bare places, 8 desatins ; sold 2 desatins for 549 roubles.

'2. Zaidovsk, 158 desatins 850 fathoms, of which there are forest area, 143 desatins 700 fathoms ; plantations by forest guards, 5 desatins 400 fathoms ; appurtenances, 8 desatins 500 fathoms ; waste, 1 desatin 1,650 fathoms.

Oak is the predominant plant; 5 desatins were sold for 857 roubles.

'3. Plosko-Samarin, 152 desatins 857 fathoms. Of this there are forest area, 94 desatins; to forest guard, 10 desatins 800 fathoms; appurtenances, 44 desatins 400 fathoms; waste, 3 desatins 1,557 fathoms; sold 5 desatins for 717 roubles.

General Remarks on the Forests of the Governments of Kherson and Ekatherinoslav.

'In the governments of Ekatherinoslav and Kherson the forests are barely 1 desatin to 100 desatins of the general area, and this is according to the statistics of the Russian empire; although more than the actual proportion it is very near it. In the survey of each government we have stated why the cipher of forests is generally higher; but if we accept the figures given, and reckon a little larger amount of forest area, then in that case the Kherson and Ekatherinoslav governments present a rare exception in Russia for their having few forests.

'Notwithstanding the proximity of the sea and the great rivers, as for instance the Dnieper, Dnester, Bug, and others, water these governments at short distances, droughts in these governments are very frequent, and with these there are frequently bad crops.

'The want of forests in these governments it is difficult to meet by means of plantations. In the governments of Ekatherinoslav, Kherson, and the northern parts of Taurida, forest plantations have made more progress than in other governments. The example given by government was not lost, many private individuals having occupied themselves with planting trees in the suburbs of Odessa and Nicolaiev. In the districts of Kherson, Bobrinetz, Ananiev, and Alexandria, plantations have been made on many tens of desatins, and there are already two estates with several hundred desatins—in the village Trikratach, 380 desatins, and in Skalevatnack, 260 desatins; and

besides, in the crown villages they have been planting with
considerable success.   But if we reckon the planting of
trees in this country from the sowing of acorns in the
neighbourhood of Nicolaiev, by the order of Prince
Potemkin, in Taurida, then it will appear that the
planting of forests has since then made little progress ; and
at the present time in the three governments, Kherson,
Ekatherinoslav, and northern parts of Taurida, not
more than 10,000 desatins have been planted.   During
the time of serf labour many occupied themselves with
this on private estates, but have given over doing so now,
commercial calculations having put a stop for many years
to the slow process of forest planting.

'In the meantime the planting of forests in this
country has brought to it a practical and very palpable
benefit.   In this respect the first place is occupied by the
small forest near Odessa, on the Peresif, planted in 1831 and
1834 on 130 desatins of salt soil, to defend Odessa from the
wind, which brought clouds of sand into the town.   This
planting was made on the idea of the former chief of
Odessa, Leoshin, and it is known by the name of the Leoshin
plantation.   Since then these plantations have increased,
and one must remark that planting of trees on the
Peresif, composed of sea drifts of sand and closely packed
shells, with a salt soil, and want of water, was difficult,
but it proved to the inhabitants of Odessa the possibility
and utility of plantations, and from that time the suburbs
of Odessa, little by little, have begun to clothe themselves
with green trees ; but they are very sickly, and short lived,
because the heat, dust, hardness of soil, want of water,
and frequent droughts, do not give the trees a possibility
to develop themselves properly, and only those trees which
are protected from all sides, at least from winds, the
drying nature of which forms the greatest danger to
vegetation, grow pretty satisfactory.

'The Leoshin plantations had the same importance as
the great Anadol plantations in the government of
Ekatherinoslav ; in both cases they were designed to prove

the possibility of making forest plantations. The difference
between them is, that the Leoshin plantations were called
for by necessities of the town, and have brought to it
a practical benefit by defending Odessa from sand carried
by the wind. Besides this, many capitalists of Odessa
were so carried away by this example that forest planting
was for some time fashionable ; they used to boast of it,
and planted trees on estates without any views as to the
income and economical significance of their planting—
there was only one object in view, the so to say climatic
influence of these plantations on the neighbouring estates :
and of the climatic influence the inhabitants of Odessa are
strongly persuaded, because the summer heats and
neighbouring *steppes* frequently remind them of this.

'Forests could counteract droughts, because their
influence on the winds in the *steppes* is very visible ; roads
and ploughed fields under the influence of wind soon dry ;
but the drying proceeds much more slowly, not only if the
soil be defended by forests, but even if it be so by burian or
high grass. When we see thousands of desatins of ploughed
land loosened for several vershocks in depth, and very
rapidly drying under the influence of strong winds, then
we have a visible indication of what immense stores of
water the soil must have in order to satisfy on one side
the rapidity of this evaporation, and on the other side the
demands of the future vegetable life. The rapid movement of the layers of atmosphere nearest the surface must
have great influence on the moisture of the soil.

'There is not much wood in the governments of
Ekatherinoslav and Kherson, compared with other governments ; but for that the demands for forest materials are
very limited, and these are used very economically. Mr
Konoplin, talking of the wood trade in Prussia, held up as
an example the economy in the use of forest materials shown
by Prussians ; but this economy in the governments of
Ekatherinoslav and Kherson is perhaps greater than that of
the Prussians : here you will not only not see any log or piece
of timber lying about, but even chips and bark remaining

from the logs in landing them is carefully collected from the mud of the Dnieper by the poor inhabitants, amongst whom there is a lively trade in this material. The Jewish population employ part of the materials received in this way for manufacturing different small goods from wood; the sawdust goes for fuel; brushwood is strictly sorted in Odessa, and it is partly employed in making baskets, which in many cases replace tubs, which are very dear in this town. With the above retail sales of firewood, in Odessa firewood is subjected to the most careful sorting, the size of the billet, thickness, state of rotteness, dampness, straightness of grain, everything is taken into account, and expressed in kopecs.

'The very construction of wooden buildings is done most economically; in most cases farmers of the middle class use in building houses of medium size, being in length 9 to 12 archines, in breath 7 to 8 archines, the following quantity of forest materials :—

' For posts in the walls, for ties between the posts, and for the beams of the ceiling, 10 pieces of 9 archine beams, 5 vershocks thick; for truss pieces, 20 pieces of 9 archine beams of the thickness of 3 vershocks; to the length beams they attach crossings of perches, and to the truss pieces split perches, 50 pieces of 9 archine beams, 2 vershocks thick.

' Further, for the making of doors and windows, about 5 beams, 9 archines long, by 4 vershocks thick; boards, 7 archines long, by 4 vershocks broad, 8 pieces; deals, 15 pieces, and some pieces of slabs, and two loads of brushwood. Neither deals or boards are used for the floor, ceiling, or roof. Beams for walls are likewise not often used. Such examples of economy will probably not frequently be met with even in Prussia.

' The variety of forest materials required in the wood trade, as well with regard to kinds as with regard to dimensions, makes it difficult to fix prices for these materials, and considering the number of materials it would be easy in the wood-yard to mix the prices. The

necessity to avoid this inconvenience has led to the
system of considerably simplifying the business—most
frequently it is usual to fix the price with timber of the
same length from the vershock of thickness, and that of
the same thickness from the archine in length; in the
same way is fixed the price for carriage by land by pro-
portioning the prices to the cubic contents; and the
weight of the materials also supplies a means of satisfactorily fixing the price for transport.

'In addition to this the price in sales by the archine or
vershock differs—one price is fixed for the archine if the
sort is from 9 to 15 archines in length, another if 18
archines, and so on; and in the same way likewise for
the vershock of thickness. Such a system might be
adopted for estimates of crown forests, because it answers,
as it were, the demand of the wood trade, and likewise
takes account of the cubic contents of different sorts, and
the cost of production of forest materials.

The Present State of the Wood Trade in the Government of Volhinia.

'In latter years the forests in the government of Volhinia
have been cut down and cleared to a much greater extent
than formerly. For this reason the wood trade of this
government is far from being in a satisfactory state; sundry
wood dealers receive very different profits from wood—
instances of receiving great profits are intermitted with
cases of great losses. Generally the wood trade in this
government requires great caution.

'The principal occasions of such a state were—the cheap
sale of forests by parties intending to take an active part
in the troubles that arose in the south-western country,
and the great demand for forest materials that ensued
soon after these sales for Prussia, in consequence of known
political events. The co-incidence of these two events
gave many persons considerable profits; these were in
many cases 75 to 150 per cent. Before this time the

fellings in the government of Volhinia were much more limited than in the governments of Kovno, Grodno, and partly Minsk; and generally the dealers in forest products were very little known in Volhinia, and did not risk to make in it great purchases. When the demand for abroad increased, then in the forests of the governments of Grodno, Kovno, and partly Minsk, so long worked, they had to make fellings in places little convenient for meeting this demand, which had a tendency to raise prices for the timber destined to be taken out of the country. This rise of prices in connection with the cheap and great sales of wood, and there being no increased demand for labour in the government of Volhinia, afforded a means to a few wood dealers well acquainted with the forests of the government of Volhinia to make great profits. But such a state of affairs did not continue long; it soon changed to the detriment of the wood dealers. In the meantime the rumours of the profits had become known, and many hastened to take part in the wood trade, or to increase the business in it. When by this means the decreasing demand on the one side, and the increased offers of sale on the other, lowered the prices, some hastened to sell, others held, and in consequence of this the different results to the dealers became extreme. Many that waited for a more propitious time lost; those that hurried to sell received less profit than in former years; and finally those who bought woods in time, and managed to make a sale before the advent of this unusual state of the wood trade, gained considerable profit. Such variations in the wood trade during the last five years has been reflected on the income of the crown forests. In 1857 the revenue from the crown forests was 11,588 roubles, which increasing gradually, attained in 1861 to 42,215 roubles for sale of timber only; and in 1862 it fell to 36,753 roubles; in 1863 to 28,591 roubles; but in 1864 it commenced to rise, and attained to 40,000 roubles; in 1865, 62,000 roubles; and in 1866, 79,000 roubles; so that two triennial periods (from 1861 to 1863, and from 1864 to

1866) present different results: in the first the yearly
decrease of almost 10,000 roubles; and in the second the
annual increase of almost 20,000 roubles, in consequence of
which the income in three years almost doubled itself.

'The gross revenue from the crown forests of the
government of Volhinia were—

		1862.	1863.	1864.	1865.	1866.
		Roubles.	Roubles.	Roubles.	Roubles.	Roubles.
From sales of timber,	-	36,753	28,591	40,624	62,070	79,299
Sold by reduced scale,	-	13,203	14,418	15,617	19,565	17,821
Gratis, - - -	-	3,613	11,023	9,813	12,425	5,941
Received for billets and interest,		649	945	1,258	1,816	2,068
Total revenue from leases and other sources,	- -	40,677	31,922	46,641	72,338	102,587

'The government of Volhinia deals principally in produce for abroad; it sends this to Riga and Prussia. The
participation of this government in the interior trade with
the Dnieper governments having a scarcity of forests is
very inconsiderable, which is principally owing to the
want of good communication; the land carriage to the
governments of Podolia and Kherson is hindered by the
dividing ridge between the basins of the rivers Pripet and
Bug, but the water communication is much more convenient to Riga and Prussia for timber, and to Warsaw for
firewood and ordinary building materials. For the Dnieper
the principal wood products of the government of Volhinia is
pitch and tar, but of both of these products comparatively
little goes. The Kiev-Balta railway, with branches on Berdichov and Volotchisk, will not have any great influence on
the sale of the principal mass of Volhinia forests, in consequence of their distance. The branch to Berdichov
will have a great influence on the most southern estates of
the Gitomir forest district, and the branch to Volotchisk, on
the south-western estates of the government of Volhinia.
The opening of these railroads will in all probability not
take place soon, and until then the sales from the forests
of Volhinia will be principally influenced by the state of
the wood trade in Warsaw, Danzig, and Riga.

Two Principal Groups of Forests in the Government of Kiev, and their bearing upon the Wood Trade.

'The crown forests of the government of Kiev are divided into two groups in respect to the wood trade, and even in wood growth, viz.,—the northern and the southern. '

'The northern group of forests in the government of Kiev consists of forest estates of the districts of Kiev, Radomisl, and Vassilkov, and forms, in respect to the wood trade, one whole with the forests of three Oster and one Tchernigov forest districts in the government of Tchernigov, the Pereiaslav forestry in the government of Poltava, and the southern part of the Retchiltza district in the government of Minsk. All this ma·s of forests is so disposed round Kiev that it can be considered as one estate, and not as separate estates in different governments. Almost in the middle of this mass of forests the rivers Pripet and Dwina fall into the Dnieper. The junction of these three important floating ways has a great influence in respect to the wood trade on these forests, because the Pripet, and so likewise the Dnieper and Dwina, before their junction cut through a great extent of forests. Many wood dealers on a small scale, collecting forest materials in places distant from one another, have no idea of the whole mass of wood prepared for floating down the Dnieper. With the opening of the navigation about Kiev the wood dealers from the rivers Pripet, Dwina, and the upper part of the Dnieper, come for the first time into communication between themselves, and here only learn for themselves the real state of the wood markets with regard to the quantity of prepared materials. The fluctuation of prices continues until the last rafts go. Many small wood dealers take the wood only to Kiev ; here the forest materials pass into other hands, and go to Ekatherinoslav, and very rarely to Kherson, because the great part of the Kherson rafts have their destination from the starting point, in order not to lose time in Kiev, and not to let slip through this the

highest water over the rapids, which facilitate the passage
over them.   And when it is necessary for the Kherson
rafts to winter then they winter principally at Kremente-
bug.

' The wholesale purchases for distant floatage have a
great influence on the sale of wood floated down for local
use ; and this in its turn is greatly dependent on the state
of trade generally on the river Dwina.   On this river are
many fabriques, and many goods are floated down on rafts.
The principal employment of these rafts is to carry goods,
in consequence of which these rafts can be sold cheaper in
Kiev than those that do not carry goods, as the first
derive considerable profit from carrying these goods.   The
number of such rafts, and the character of the forest
materials composing them, depends on the demand for
fabrique goods, and therefore vary from year to year.

' Under the influence of such different circumstances
the prices for forest materials are fixed between the
mouths of the rivers Dwina and Pripet, and principally at
Kiev ; and the state of these prices has an influence on
the sale of forest materials from the crown estates concentrated, as stated above, around Kiev.

' Besides the effect of these circumstances on the sale of
forest materials, from the estates of the northern group, a
great influence comes from a number of works going on near
the estates of the northern group, and principally at Kiev.
When there is scarcity of work the rural population
occupy themselves with the sale of materials that cost
them nothing, taken from the private, and perhaps even from
the neighbouring crown estates.   The result is sometimes
an increase in the quantity of forest materials left in the
hands of the more considerable dealers, and these remains,
particularly those of firewood and small building materials,
are not without influence on the price of forest materials
of the next season.

' Still greater unity to the forest estates round Kiev is
given by the Kiev-Balta and Kiev-Koursk railways.
These roads give a facility for the disposal of firewood for

use of the railways themselves, and for the sale of building materials in the districts possessing little wood in the governments of Tchernigov, Kiev, and Podolia.

' The southern group of forests of the Kiev government is at a considerable distance from the northern, and consists of the forest districts of Svenigorod, Tchiguirin, and Tcherkazy. This group of forests, in regard to wood trade, is similar to the estates of Novornigorod and Krilov forest districts in the government of Kherson, and Krementchug, government of Poltava. All the estates of these forest districts are not far from one another, and make, as it were, one whole area, at a considerable distance as well from Kherson as from Kiev and Poltava, where the principal administrations over the forests are situated.

' This group of forests being near the parts of the governments of Kiev, Kherson, Ekatherinoslav, and Poltava, possessing little wood, competes successfully with the floated wood, with the exception only of those estates (principally near Krementchug) which are very near the Dnieper, almost on the shore.

' In this group of forests the great dealers cannot concentrate in their own hands the forest materials prepared by the small dealers, but must compete with them. This circumstance is influenced principally by the great demand for forest materials by the local consumers, small and great, particularly sugar refiners. The Balta-Krementchug railway will exercise a great influence on prices, as it will afford a possibility of taking these materials to the government of Kherson, between Elizabethgrad, Olvicopol, Bobrinetz, and Voznessensk, where at present there is great want of forest materials in consequence of the difficulty of conveyance. Such expectation of forest trade in these places has occasioned already in 1867 great sales from crown estates : for 25,000 pine trees 177,000 roubles was paid. Such prices did not exist before.

' Under these circumstances, the forests of the Kiev

government, being on the boundary of the forest belt of the Dnieper basin, promise a considerable increase of revenue from forests. The following data may give an idea of the importance of the wood trade on the crown estates of this government :—

	1862.	1863.	1864.	1865.	1866.
	Roubles.	Roubles.	Roubles.	Roubles.	Roubles.
Sold, · · · ·	39,288	29,727	35,800	59,720	86,662
Sold at a reduced tax, ·	918	960	860	1,301	3,156
Given gratis, · · ·	4,532	3,544	6,774	8,677	20,116

'The northern group of forests of the Kiev government comprise the following—

	Forest area. Desatins.	On Lease. Des.	Belonging to Foresters. Des.	Appurten-ances. Des.	Waste. Des.
Kiev, 1st District, -	35,656	275	66	150	1,629
,, 2nd, · ·	21,094	60	84	233	397
Radonisk, 1st, ·	18,154	—	44	214	1,202
,, 2nd, ·	27,512	—	30	—	324
,, 3rd, ·	23,400	—	2	—	8,019

The southern group—

Tchigivirin District,	15,632	1	158	656	13,812
Tcherkaszy, · ·	28,204	—	160	68	539
,, 2nd, ·	10,425	—	13	11	114
Svenigorod, · ·	16,704	558	30	--	—

'Besides these forest districts there is also the Berdicher, which by its position, and conditions of sales, forms part of the Jitomir forest district governments of Volhinia, and is not a separate forest district of the government of Kiev.'

CHAPTER IV.

FOREST EXPLOITATION.

In the account given of forests and forest management in the valley of the Dnieper, allusion is made to certain forests being organised in contradistinction to others which are not so ; and I have stated that in bringing under consideration the forestry of Lithuania I was influenced by the circumstance that this would supply an opportunity of bringing under notice the forest management of Russia, as seen in the administration of forests comprised in the Imperial domains.

The state forests of Russia proper, exclusive of appanages, or forests included in lands set apart for the maintenance or pleasure of members of the reigning family, are under the administeration of the Minister of Imperial domains. In each government is an inspector of forests, with numerous subordinates, who are intrusted with the management of the forests. These occupy a good social position ; all the superior subordinates of the inspector are educated men who have passed with credit through the professional instruction and training prescribed, and supplied at the schools of forestry. The forty-two governments are grouped according to their geographical position in eight forest divisions, and there is published annually a report entitled *Otchet po Laesnomu Upravleniou Ministerstva Gocydarstven-neech Eemustchestva* in which, under five chapters, are supplied :—

1. Statements relative to the extent and contents of the forests under the administration.

2. Statements relative to the organisation for the management of these forests.

3. Statistics and economics of these forests and of quit-rent places connected with them.

4. Statements of means employed for obtaining revenue from them, &c.

5. Financial accounts.

The laws and regulations relative to the conservation and exploitation of the state forests have been codified, and published under the title *Ystav Laesnoi;* and at irregular intervals of one, two, three, or more years there are issued alterations and additions to the code, or revisions of the code, stating what laws or regulations have been confirmed, abrogated, or altered. In Lithuania, as elsewhere throughout Russia—with the exception of the northern forest zone, in which the demand for timber for exportation in the extreme north, and the demand for wood in large quantities for mining operations, and perhaps the government of Tula, where there is a like demand for manufactories of metal wares, has given special characters to forest exploitation—the principal demand for forest produce is to supply what is required within the district, and in accessible districts within the empire, for building purposes, carpentry, and fuel. In meeting this demand the government comes into competition with private proprietors of forests. These are under great inducements to sell expeditiously all that their forests produce; but the state can afford to delay felling more than is necessary to meet existing requirements beyond what can be so met; and by acting on this principle the subsequent well-being of the community, in so far as this is involved in the conservation of the forests, can be secured without interference with private property. It is extensively held by students of forest science, that it is only in forests belonging to the state that the full benefit of forest possessions can be secured to a country. The lifetime of a man is not yet equal to the life of a tree; and in many cases it is only by allowing a tree to attain its maturity that the best results can be obtained. A man may be willing to plant and to incur trouble and expense in the maintenance and conservation of a wood in view of the prospective good which may be reaped from this by his children or even his children's children ; but he may feel less enthusiastic if the benefit is only to be reaped by his great-grand-children ; and any one of a hundred things may occur before these be born or come of age, to render it expedient in connection with personal interests to fell and sell, and not replant. But a nation never dies ; according to many all forests, but beyond question all state forests, are the property of the nation in its entirety—past, present, and to come, of which each passing generation has the usufruct, like the holder of an entailed estate, but is bound in justice to pass it on in like good condition as that in which they found it, or with compensating advantages for what in view of national interests may be destroyed by them; and as the state never dies, the state can therefore afford to delay felling till the full benefit of the possession has been obtained. Many of the students of forest science hold, and I hold with them, that it may be all very well to encourage private planting, and to do so in every way compatible with the common good ; and that probably only good will result from such private enterprise ; but that in order to ensure a continuous supply of forest produce of national growth it is necessary to have extensive state forests under wise administration and scientific management : and in accordance with this is the administration and management of forests here, and throughout the central governments of Russia.

The general impression produced on my mind by all I have learned is that the exploitation is on what is known in France as *La Methode à Tire et Aire*—a rough division of the forests into sections to be successfully exploited in successive periods—with the occasional practice of *Jardinage* or felling of trees selected as suitable for some purpose designed, when this is deemed expedient. But all this is done with a general tendency to introduce, or at least prepare for exploitation, according to what is known in Poland as the scientific method of exploitation ; that known in Germany as *Die Fachwerke Method ;* and in France as *La Methode des Compartiments,* in regard to which details have

O

been given in a preceding chapter, relative to forest exploitation in Poland.   And the policy seems to me to be to entrust the work to educated foresters with general instructions, but with great freedom of action, so as to secure the application to the details of the principles involved, leaving them free in the determination of the application of these, as the medical practitioner is left in determining the application of the principles of his profession, to be made in the case of any and every patient under his care.

Of the views entertained in Russia in regard to the different methods of exploitation, which have engaged the attention of students of forest science, I have given in *Forests and Forestry of Northern Russia and Lands Beyond* (pp. 101-108), a translation of a statement of these by M. Werekha, in a *Notice sur les Forêts et leur Products, &c.*, prepared by a special commission charged with the collection of products of the forests, and of rural industry, for the International Exhibition at Vienna in 1863.

The following is an account of the transport and preparation of timber on the rivers Dnieper and Berezina, published in the *Transactions of the Scottish Arboricultural Society* as an abridgement of an article on the subject in the *Timber Trades' Journal :—*

'The business is done here on no small scale ; the amount of wood yearly floated down on these two rivers is immense.   From the moment, in the early spring, when the ice melts and the rivers rise some ten to sixteen feet above their usual level, we see the rafts coming down in succeeding masses

'Like all business in this part of Russia, the wood trade is in the hands of the Jews.   Owners of estates and forests sell part of their wood to them, and they know how to make the best of everything that comes into their hands.   Winter begins here generally in November.   In September, when the peasants have to pay their taxes, contracts are made with them, when generally those who are living together in small villages agree, and bind themselves, against an advance, to cut and drive a certain quantity of wood down to the banks of the rivers. These advances are sometimes a third or one-half of the amount they can expect to earn during the course of the winter, but when the agreement is signed by the elected members of the court of the village, or the Starosta and the Uraduisk, as they are called here, there is hardly any risk of loss.

'In November the peasants come to the woods, each with one or two, sometimes three, of their small pony-like horses, in the last case called *troskas*, collecting together often as many as 200 to 300 horses from one village. A sort of abode for the winter is then erected in the woods, built of small poles, earth, and straw, which reminds one more than anything of the huts of the Esquimaux and the Laplanders. The building is made in the following manner: earth is thrown up so as to form a round flat cake, 1 foot high, and 12 to 15 feet in diameter. On this platform poles are placed in the shape of a sugar-loaf. On the poles, at the top of which is a little hole for the smoke to escape, is laid straw, and on the straw earth and sand; and in the side of this extraordinary Russian mud-house there is an opening made for ingress and egress. When ready, 12 to 15 men make it their home for the winter. Furs, rags, and little boxes for provisions are placed all around, and the fire, composed of large logs, with the large saucepan, in the middle. When the work of the day is over, the workmen seat themselves each on his place round the flaming fire, on which the soup, composed of meat, cabbage, and onions, boils; this is the time to see the Russian peasant, and to hear his monotonous chant, reminding one of the inhabitants of some wild country. As for him, he has no delicate nerves, and his smelling organs seem to enjoy the smoky air as it becomes heavier and thicker; he puts his rags round him, stretches himself out on the sandy ground, and, unmindful of storm or cold, sleeps the sleep of the innocent till the morning light, which wakes him up and reminds him it

is time to put the primitive harness, generally made of
rags and rope, on his poor half-starved horses, and go to
his day's work again.

'The cutting and driving is done in this way.   Every
peasant has his axe ; he fells his tree, clears off all knots
and branches, lays it on his sledge, and drives it to the
river.   Thus millions of trees are brought down these two
rivers in the course of the winter.

'The kind of wood grown here is a sort of redwood fir,
sometimes also whitewood.   The fir tree grows very fast.
A fir tree which requires 120 years and more to ripen in
the north of Europe matures here in 80 years.   It is,
however, coarse, sometimes sappy, and contains a mass of
resin and other matters, which makes the smallest knot
of a bright red colour.   To see the trees standing in the
forest is a fine sight when they are straight, and grown
high without branches ; but cut them down and the charm
is gone.   Masses of timber, quantities of firewood, chiefly
birch, elm, alder, beech, and other kinds of wood, are
forwarded down the river during the entire spring and
summer ; most of it to the Black Sea, but a small part of
it is taken up by river to the Baltic.   Most of the logs
are formed into large rafts.   A hole is made in the end
of each log, and they are tied together with bands of
willow.   This is a clumsy and an expensive way of constructing rafts, besides wasting two to three feet of each
log.   Two or three rafts are then tied together with
willow bands, a little wooden hut is erected, sometimes
hardly larger than a dog kennel, on which is a tiny pole
with a bit of red or blue cloth as a flag.   The three or
four men who are in charge of this raft make it their
home for the six or seven weeks (sometimes more) that
they are on their way to the Black Sea.

'Another way of transporting different kinds of wood
down the rivers is in large lighters, sometimes called
Berliner, sometimes Barkar.   The former are very
strongly built, but the latter are of enormous planks 60 to
80 feet long, large enough to load 200 to 300 standards,

and only built for the one voyage down to the Black Sea, where they are taken to pieces and sold. I had an opportunity of seeing both of these kinds of lighters built last winter. It was a queer sight to see three such Berliners building, each large enough to hold about 120 standards. There can be no doubt that in like manner the men-of-war were built some hundreds of years ago, when many battles were fought here between the Russians and the wild tribes from the east and south of Russia. The boards used were 5 to 6 inches thick, 15 to 20 inches wide, 50 to 60 feet long, hand-sawn out of one block of the finest trees found in the woods. Thousands of the most beautiful oaks were cut down and made into lighters, each being equal in value to that of a nice sized schooner. Still more wonderful was the building of the barque, which is a sort of Noah's Ark as to size. How they got this enormous structure to hold together and to keep tight is not easy to understand, more especially as it was built to pass the cataracts between Kremenchuek and Kherson. Most of the timber is being sent down to Kherson, Nikolaiev, and Odessa; a small part of it is sold on the way at Kiev and some other places. In Kherson are large sawmills, where many of the logs are transformed into deals and boards; others are shipped from Nikolaiev and Odessa in the form of square timber.

'There are some sawmills in this part of Russia also, but they are as old-fashioned as everything else, where some thousands of logs yearly are sawn into large boards, mostly all of one size, and sold at so much a piece at Kiev.

'In this part of Russia there exists a decided feeling against foreigners; and with the dim idea they have of right and wrong, they consider it their duty to persecute strangers as much as lies in their power.

'In the end of 1882 a wood-exporting firm in Finland made an agreement with a Count v. M., in St. Peters-burg, who was the owner of a large estate with extensive forests in this neighbourhood, to take out the value of the

woods for joint account.   The forests contained about a
million of trees, ripe for cutting, and these were to be
made into money in as short a time as possible.   Plans
were made ; a sawmill with six frames, and a planing-mill
were to be built, and 80,000 trees were ordered to be
felled the first year.   The trees were felled, the sawmill
was built, workmen were collected from Sweden, Finland,
and Riga.   Last summer the sawmill was so far ready
that sawing began, when the firm in Finland unexpectedly fell into difficulties.   Money was not sent to pay the
workmen.   Some time after, the firm in Finland became
bankrupt, and the owner left for America.   The Count v.
M. stopped payment in the real sense of the word, and
there the poor workmen were left with their wives and
children in utter want of money, in an exceedingly
dangerous climate, where fever and illness came more
regularly than the daily bread, without means to buy
medicine, and without a medical man to attend them.
Death visited them through typhus, and they had to bury
their dead themselves.

'Te detail the intrigues, the unfulfilled promises, and
the mean behaviour of the Russians against these poor
people, would be of no use.   Suffice it to say, that by
their common efforts they got over the first part of the
winter, and through the help of the Swedish Ambassador,
and the Finnish authorities in St. Petersburg, they have
been sent home to their respective countries.   The business is entirely wound up, and the very fine sawmill, with
its first-rate machinery and every new improvement, is
waiting for a new owner, who may have sufficient means
to make himself independent of Russian intrigues, and be
able to continue a business which began hopefully a little
more than a year ago.'

Kherson, the capital of the government of that name,
was founded in 1778, and soon became a port for vessels
from all countries of Europe.   It is 57 miles from the
Black Sea, and 92 E.N.E. from Odessa, on the right bank

of the Limari, an immense embouchure of the Dnieper, which is here four miles broad, where, when as is frequently the case, its numerous shoals are covered with water, but when the shoals are exposed, the breadth of the river itself is not more than one verst, or two-thirds of a mile.

It is an emporium for the equipment and armament of the fleet of the Black Sea, timber being brought by the Dnieper both for its own supply and that of Nicolaiev and Odessa. There is a fine basin cut out of the limestone rock. During the spring flood of the river vessels built here can be transported to the Black Sea upon 'camels,' as they are called, and much of the produce of the interior is brought here, and taken to Odessa in lighters.

The St. Petersburg *Vedomosti* gives particulars of the reclamation of a vast track lying between the rivers Dnieper, Pripet, Beresina, and Ptitshja, known as the Polessje region, which has been hitherto useless and in great part impassable. The works began in the year 1874, and by the end of last year a canal system of about 1,695 versts (1,130 miles) had been completed, which had already drained 1,141,000 desatins (over 5000 square miles.) Nearly one-sixth of this vast track, which had previously been an impenetrable morass, has been changed into meadow land. An area of over 1,250 square miles of forest, which was totally useless, being traversed by a network of swamps, has been thoroughly drained, the lines of swamp being cleared, deepened, and converted into drainage canals, which, unfortunately, can have only a very slight fall. Another large piece of forest, over 750 square miles in extent, hitherto practically inaccessible, has been opened up by canals and made available for useful purposes; and the remainder, amounting to about 740,000 desatins (over 2,800 square miles) has been drained and brought into a condition fit for cultivation or pasturage. The work is said to have been executed at an annual expenditure of 265,000 roubles, or a little less than £40,000.

Of the dexterity of the Lithuanians in what I may call woodcraft is thus incidentally alluded to by Mr Anderson in his account of *Seven Months' Residence in Russian Poland in* 1863, which has been already cited :—

'The birch tree is, to the Polish peasant, the most useful tree of the forest. His furniture, cart, plough—in fact, all his agricultural and garden tools—are made of this wood. It seems hard and strong enough for all purposes, and serves even for the teeth of his harrow, and for the lower part of his spade, as well as for its handle. He constructs, also, out of the same material, long forks, with which he contrives to throw up to a great height the sheaves of corn gathered into their barns. In this work two men stand with their backs to the place where the sheaves are to be stored ; they then stick both their forks into the same sheaf, and, upon one of them giving a grunt, up it goes, flying over their heads, to its destination.' And again :—

'Considerable ingenuity is sometimes displayed by the peasants in the execution of their work. I once saw a man, who had invented a kind of turning-lathe, in order that he might rapidly finish the nave of a cart-wheel upon which he was engaged. He had fixed the piece of wood on which he was at work upon two iron pivots. He then twisted a rope twice round the piece of wood ; attached one end of the rope to a strong birch sapling which he had fastened in the ceiling ; and, in a loop at the other end, he put his foot. He then set in motion the wood, upon which he was at work ; and the spring, given by the sapling, acted as a lathe. He had in his hand a stick, with a strong crescent-shaped piece of iron fixed to it ; and with this he worked away, just as if he had the best turning-lathe and chisel in the world.'

CHAPTER V.

THE JEWISH POPULATION.

The black sheep, the *bete noir*, of the Lithuanian patriot, and of the enthusiastic forest conservator in Lithuania, is the Jew. I hold in high estimation the nation ' to whom pertaineth the adoption, and the glory, and the covenants, and the giving of the law, and the service of God, and and the promises; whose are the fathers, and of whom as concerning the flesh, Christ came.' I have met with noble men and women among them. Even amongst those of them who do not consider that Jesus of Nazareth was the anointed King, whose coming was, and is still, expected by the people, there are men, and at least one large community, manifesting a spirit such as when seen amongst Christians is held in high esteem by the more devout: but all are not such; and of them, equally with the British and the Anglo-Americans, there are worshippers of the Mighty Dollar; of them, as of the other nations named, it may be said they are like the prophet's figs—the good are very good, but the bad are very bad.*

* In the beginning of 1840, at St. Petersburg, I made the acquaintance of Pastor Boerling, a clergyman of the Lutheran church, and himself a descendant of Israel, who stated to me, amongst other things, that he was stationed as a missionary for many years at Schloss, a town in Poland, which is inhabited chiefly by Jews. When he first went there he saw no opening for usefulness ; and after a little time he began to fear that he had run unsent. But the cholera soon broke out in the place, and all the medical men fled ; he then concluded that he had been sent thither of God—for a previous residence in several towns of Asia, while the cholera prevailed in these places, had made him acquainted with the most approved methods of treating the sufferers, and now the people implored his aid. He cheerfully attended the sick, and soon gained their affections. From that time their houses were open to him, and he was invited to all their entertainments and feasts.

On one occasion he was present at a marriage feast, when, according to custom, all the guests presented gifts to the newly married pair. He had just received from London a few copies of a 12mo edition of the Hebrew Old and New Testament bound together, and he presented them with one of these. It was gratefully received, and at the close of the feast, when the bridegroom held up the different presents, and announced the names of the giver of each, exhibiting the Bible last, he said, ' But see what our friend the missionary has given us—the Scriptures ! This I value more highly than silver or gold !

In passing through Lithuania, they seem to swarm so as to suggest the illustration of locusts eating up every green thing. No one seems to have a good word for them; while every one seems to have something to tell against them. On learning a little more of facts than can be gathered only from finding the platform of a railway station

---

The young man took the Bible regularly to the synagogue when he went to worship. The Reader, observing this, demanded of him how he dared to bring the Christian book into the synagogue. He replied, that he had read it through, and found nothing ungodly in it; and that he must and would read it. Many of the other Jews then applied for copies, with which they were supplied; and the desire for instruction became so great that the inhabitants of the town requested the missionary to organise a school for the instruction of the young. He complied with their request, organising one for the instruction of boys under his own superintendence, and another for girls under the superintendence of his wife

He met with opposition from quarters whence he had least reason to expect it, but the great body of the Jews encouraged him; and after some time a Jew of considerable learning and influence came to him and said, 'One or other of us must leave this town. If you don't go, I go; for if things go on thus, my children also will be taught to read, and to read the books of the Christians.'

He also mentioned that he was appointed at one time to labour in Upper Silesia. He went thither, and on approaching one town, the first he entered, he was informed that all the inhabitants were Jews, but that he would have no opportunity of prosecuting missionary labour there, for they were all rich and wanted nothing. On entering the town he was soon convinced of the correctness of the information he had received; but as a few Christian Jews resided there he resolved to spend a few days in intercourse with them. It was then Friday, and on the following day he went to the synagogue. Several of the Jews assembled there, observing him to be a stranger, welcomed him with the usual salutation of, 'Peace be with you!' When, however, they observed that during the prayer which was offered he stood devoutly and still, instead of looking about as did others, they whispered aloud, 'He is not a Jew but a missionary, for all the missionaries pray so.'

What were the consequences? In the course of the day many of the Jews visited his apartment for conversation concerning Christianity; and they spent the time not in disputation as at other places, but in calm and dispassionate comparison of the Old Testament prophecies, with the history of Jesus of Nazareth recorded in the Gospels! In the evening *six* Jews, whose heads were silvered with age, waited upon him, and almost abjured him to tell them what had convinced him of the truth of Christianity; and they too spent their visit in a calm and apparently dispassionate examination of the attestations of the Messiah.

He assured me that ten times the number of missionaries now labouring in Poland and Silesia might find full scope for their energies in cultivating that extensive and hopeful field. The opinion prevails that the Jews present a hopeless field for missionary culture, but there are many things leading us to a contrary conclusion.

God hath not cast off his people, if, with the Apostle, we believe that God is no respecter of persons, but in every nation he that feareth Him and worketh righteousness is accepted with him; and if we search amongst the Jewish people, we may find many like their fathers, who bowed not the knee to Baal; many like the godly Jews of former days—men like Simeon, 'just and devout, waiting for the consolation of Israel.'

I felt much interested in a description given to me by Pastor Boerling, of one of his acquaintances, an aged Rabbi, who, like Anna the prophetess, the daughter of Phanuel, of the tribe of Aser, departed not from the temple, 'but served God with fastings and prayers night and day.' Regularly at the hour of midnight was that aged patriarch to be found in the synagogue making confession and supplication unto God. He was accidentally overheard on one occasion by Mr Boerling, and he repeated to me the prayer, which a retentive memory enabled him to recall. While I listened to it, I thought I saw before me Daniel when he set his face unto the Lord God, to seek by prayer and supplications, with fasting, and sackcloth, and ashes. The spirit was the

thronged with Jews in their gabardines ready to turn an
honest penny by exchanging the foreign money of travellers for the coin of the realm, and trying thus while
serving the traveller to make their plack a bawbee as
many a Scotsman does by honest trading, the stranger
may find reason to conclude that there is nothing surpri-

same, the expressions similiar to those which characterised the prayer presented by
that prophet, and recorded in the 9th chapter of the book which bears his name :—

'O Lord, the great and dreadful God, keeping the covenant and mercy to them that
love him, and to them that keep his commandments ; we have sinned and have committed
iniquity, and have done wickedly, and have rebelled, even by departing from thy pre-
cepts and from thy judgments ; neither have we hearkened unto thy servants the
prophets, which spake in thy name to our kings, our princes, and our fathers, and to
all the people of the land. . . O Lord, to us belongeth confusion of face, to our kings, to
our princes, and to our fathers, because we have sinned against thee. . . O my God,
incline thine ear, and hear ; open thine eyes, and behold our desolations, and the city
which is called by thy name: for we do not present our supplications before thee for
our righteousness, but for thy great mercies. O Lord, hear ; O Lord, forgive ; O Lord,
hearken and do ; defer not, for thine own sake, O my God ; for thy city and thy people
are called by thy name.'

This Rabbi led an abstemious life. On one occasion, when offered a little wine he
declined. In a short but thrilling reply (to which I cannot do justice in a translation),
he stated his reasons for acting thus :—' I read,' said he, ' that wine makes glad the
heart of man ; and I—can I be joyful while the city of the Lord is trampled under foot ?
Can I be joyful while the name of Jehovah is blasphemed? Can I be joyful while the
people of God, having turned their back upon the Lord, are weltering in sin ?' Is
not this the spirit expressed by the Psalmist,—' If I forget thee, O Jerusalem, let my
right hand forget her cunning. If I do not remember thee, let my tongue cleave unto
the roof of my mouth : if I prefer not Jerusalem above my chief joy.'
On another occasion he slipped away from a marriage feast at which he had been
present. He was soon missed ; and one and another of the guests exclaimed at once,
' Where is the Rabbi?' A search was made, but nowhere could he be found. At
length some one inquired, ' Have you been to the synagogue?' The parents of the
bridegroom and bride caught at the suggestion—they hastened thither, and there they
found him in the dark, engaged in prayer. They entreated him to rejoin the party and to
bless the youthful couple with his presence. He replied, ' No, I cannot. You are joyful as
is befitting the occasion of your meeting, but my heart is sad—sad ; and, when I think
of the condition of my people.' They still urged him ; when, to meet their wishes, he
consented to rejoin the party on the condition that all music should be laid aside. A
marriage party without music was an incident almost unknown amongst the Jews ; but
such was the attachment of his flock to the Rabbi, that the concession was made at
once. And on his rejoining the party, marked attention was given to several addresses
which he delivered, in the course of the evening, on the sins to which they and their
nation were addicted.
Religion is the same in all, however different may be its manifestations in different
circumstances ; and I was informed that similar manifestations of its influence were not
uncommon amongst the more humble of the Rabbies.

There was at that time a very prevalent expectation that the Messiah would appear
in the course of that year. The expectation was founded on calculations made by many
of the Talmudists, from data drawn from prophecies in the Old Testament Scriptures ;
and I was told of one learned Talmudist, who had declared that if the Messiah did not
appear in the course of that year, they were shut up to the conclusion that he must
have already come ; and if so, that Jesus of Nazareth must have been he. I have had
no opportunity of learning the effects of the disappointment which followed this
expectation.
Amongst the more learned of the Jews in those regions, I have reason to believe
there were many who were not satisfied with Judaism. I made the acquaintance of one

sing in the fact that they are everywhere spoken against;
but that those who have suffered from what they are
pleased to call their extortionate dealings, like the
innocent fleeced lambs in Britain who sow not neither do
they spin, but borrow from the Jews, have only themselves to blame for the calamities which have come upon

such,—Dr Levaison, a learned Rabbi, who was profoundly versed in the Talmud, but
found in it no satisfaction. While enquiring after the truth at one of the universities
of Germany, he became acquainted with a distinguished professor, whose theological
sentiments have secured for him a *soubriquet* importing that he is a personification of
Pagan philosophy   He gradually imbibed his sentiments, and in proportion as he did
so he had to give up his Talmudical views, but he still felt that more was necessary
to enable him satisfactorily to account for all the phenomena with which he was
acquainted. In this state of mind he met with a priest of the Greek church, who was
in the suite of a Russian ambassador at one of the German courts. He, carefully
distinguishing betwixt ceremonies devised by man and truths revealed by God, directed
his attention to the doctrines generally received as evangelical, and convinced him of
the truth of Christianity. Not having met with evangelical Christians amongst Protestants, he came to St. Petersburg in the hope of there hearing more perfectly the
principles of the religion he had embraced.
   I endeavoured to ascertain the prevalent opinions of the Jews in regard to the nature
and character of the Messiah, and found that of the Talmudists, almost all expected
him to be only a man ; among the Cabbalists, many expected that he would be divine ;
but by many of the Jews it was expected that there would be two Messiahs,—one who
has probably appeared already, in whom was to be, and has been, fulfilled the predictions contained in the 53rd chapter of Isaiah ; and another who is to reign for ever. The
former, as might have been expected, lived unknown ; but there is more than one
individual known to Jewish history whose life is supposed to fulfil what was foretold.
None, however, excepting Christian Jews, appear to consider that Jesus of Nazareth
was he. This is not wonderful, as few have access to the New Testament ; and there is
amongst them a distorted history of his life, which is calculated to hold him up to the
ridicule, contempt, and execration of the nation. With regard to that Messiah, I found
it believed that his death would be as a *sacrifice* for the sins of his people, and not
merely an *effect* brought about, directly or indirectly, by the wickedness of the nation.
   There is a very interesting body of Jews living in the Crimea, known by the name of
Karites or Caraites, and sometimes called Tartar Jews, in consequence of their speaking
the Tartar language. These men long ago rejected the Talmud, and for several genera-
tions have continued to regulate their sentiments and conduct by the Scriptures of the
Old Testament alone. I often heard of them while in Russia, and invariably on
inquiry received a favourable report of their conduct and behaviour. Many of them
appear to be spiritually-minded men ; but they were hated by other Jews, among whom
there was a trite saying expressive of their hatred and contempt, to this effect—' If a
Christian be drowning, take a Karite and make his body a bridge by which to save
him.'   But I have never heard of their rendering railing for railing. The designation
generally given by them to the other Jews, when speaking of their theological difference
is, ' Our brethren of the Talmud.' They had amongst them copies of the New Testament, which they considered a record of the life and doctrines of a godly Jew and his
disciples, and they manifested to Christians no objection dispassionately to discuss
the question of his Messiahship.
   From observation, and from intercourse with Christian Jews who have laboured
amongst their brethren, I am persuaded that the conversion of the Jews to Christianity
has been greatly hindered by the following circumstances :—
   1. Both Jews and Gentiles have fostered the notion that a Jew must necessarily
forego his nationality on embracing Christianity. It may be true that they who are like
Abraham are the children of Abraham ; but he who is a lineal descendent of that
patriarch never can cease to be such on abandoning 'vain conversations received by
tradition from the fathers.' The Apostle of the Gentiles, in common with other
Apostles,—and, I may add, in common with their Master,—was a *Christian Jew.*
   2. Jews have seldom an opportunity of witnessing the effects of Christianity in

them. The landholding community complain that the lands are passing away from them to the Jews. They are, because the former have preferred borrowing money from the Jews to enable them to live in a certain style, to curtailing their display, and thus reducing their expenditure, or to labour working with their own hands

---

'converting the soul.' They consequently form their opinion of Christianity from the conduct of men who are only nominally Christian. If they have never seen what they consider the beauty of holiness in Christians, and if all that they do see, and hear, tends to confirm their belief that Christians are utterly devoid of true religion, their prejudices against Christianity must become very strong.   We accordingly find them frequently employing the term Christian as synonymous with *blackguard*. They need, therefore, 'living epistles' to teach them, ' without the word,' that the Gospel ' is the power of God unto salvation to every one that believeth.'

3. From what they see and know of the ecclesiastical creeds and ceremonies of Christians, they consider them as polytheists and idolators, tritheists, and worshippers of saints, with dressed up representations of the virgin, and representations of God, which they consider blasphemous, as well as grotesque, and where such are not made use of, with conceptions of God scarcely less grotesque, while they have been taught to hold that God is a spirit, whom no man hath seen or can see.

4. Their usual criterion of learning is acquaintance with the Talmud. To this Christians attach no importance, and know little or nothing about it, and they are consequently despised. As in the days of our Lord so now, they make the commandment of God of none effect by their tradition. To the Jews it was commanded—' When ye reap the harvest of your land, ye shall not wholly reap the *corners* of thy field ; thou shalt leave them for the poor of thy people.' Upon this command, there are raised such questions as these :—How much must be left, if the fie'd be *four square?* How much, if it be triangular ? How much, and in what form, if it be semicircular ? How much, in what form, and where, if it be circular ?

In listening to a Jew expatiating on such subjects, one is forcibly reminded of the saying of our Lord,—' Woe unto you scribes and Pharisees, hypocrites ! for ye pay tithe of mint, and anise, and cummin, and ye neglect the weightier matters of the law,— judgement, mercy, and faith.' Such questions are discussed in the Talmud, and the first desire of an ambitious youth amongst the Jews is to study the Talmud. An acquaintance with several of the sciences is necessary to success ; and in general the student devotes himself to the study of these with the closest application, that he may afterwards overcome the difficulties to be encountered in his subsequent progress.

They appear to have a passion for such pursuits ; even boys at school challenge each other to a trial of skill in expounding the Talmud. In such cases they go to the Rabbi, and inform him of their design ; he then appoints them a passage, and they seat themselves at the extremities of the room, or in different apartments, to perform their task. In a given time they each produce a written exposition of the passage prescribed. These are submitted to the Rabbi, and the contest is determined by his decision on their respective merits.

It occasionally happens, when the children of wealthy Jews marry, that the father of the bridegroom challenges the father of the bride to support the newly married pair and their family for *twenty years*, on some other term of years, on condition of his doing the same. If the challenge be accepted, contracts are executed, and the young man generally devotes himself with close application to the study of the Talmud. If his success be considerable, his friends boast of his achievements, and congratulate themselves saying, ' Aye, he'll be a Rabbi yet !'

To attain this dignity it is necessary in some provinces to go through a protracted course of severe study. It is rarely the case that this can be completed before the student has reached his *thirtieth* year. If it be accomplished at an earlier age, the hair of the student, prematurely grey, generally testifies to his mental effort.

It does not appear to be avarice, or ambition, or the desire of usefulness, which alone prompts to the laborious and self-devoting life of a student of the Talmud. Combined with one or more of these motives, is the hope of having made some attainment whereof they may glory before God. ' They have a zeal of God, but not according to

the thing which is good; in which case they themselves
might have had to give to him that needeth. And the peasantry complain that they are impoverished, they know
not how, while their Jewish neighbours have enough and
to spare. It is so in part, if not entirely, through their
wasting their earnings upon intoxicating drinks which,
like many of Anglo-Saxon descent living amongst
people easily tempted to indulge in such stimulants, the
Jews who have invested their capital in suitable
premises are willing, and more than willing, to sell to
them; but which they need not buy or consume unless
they choose. Others, influenced by patriotism, or by
philanthrophy, or by loss coming upon them indirectly
from the evil, may have some right to complain; but the
immediate victims have none—they are reaping the
rewards of their own doings.

Of the extent to which the fathers and mothers, wives
and children, and brothers and sisters, of drunken
peasants have occasion to complain of the drink traffic,
and of the drink traffic carried on by the Jews in Russia,
some idea may be formed from facts stated by Madame
Novikoff in an article in the *Nineteenth Century*, of
September 1882, based on statements in a valuable work
entitled *L'Empire des Tzar, et les Russes*, by M. Anatole
Leroy-Beaulieu.

From the sixteenth century onwards efforts have been
made by patriotic Russians to restrain the indulgence of
the people in intoxicating drink. It is within my personal
knowledge that not a little was done by the Emperor
Nicholas. More was done in more propitious circumstances by his successor, the emancipator of the serfs.
From a tabulated statement in the *Novoie Vremia* it

---

knowledge. For they, being ignorant of God's righteousness, and going about to establish
their own righteousness, have not submitted themselves unto the righteousness of God,'
said Paul of his brethren in his day, and so may seem to devout Christians, and many
of the Jews in the present. The zeal of God is there, and in this it makes itself seen.

It seems, then, to be most desirable that some, at least, of those who devote themselves to labour amongst the Jews, should be prepared to cope with the most learned in
the discussion of the most subtile of Talmudical speculations, otherwise contempt for
the intellectual attainments of the missionary may prevent an attentive consideration
being given to the doctrines which he teaches in the name of Jesus.

appears that from 1863 to 1881 the number of drink shops was reduced by degrees from 257,531 to 146,000. A spontaneous movement, with details of which I was furnished at the time, led to the destruction by the peasants of many *kabaks*, or drinking shops, in view of their emancipation, for enjoying the full benefit of which they maintained drunkenness must cease. Previously the Baron had to support the aged serf ruined by drink, now it would fall upon his family and neighbours to do so ; and a clear head and a firm hand they said would be needed to enable them to make the most of their anticipated freedom. Thus they reasoned, and upon these views they acted somewhat riotously. Madame Novikoff writes :—

' After the death of the late Emperor the movement against drunkenness suddenly reappeared even stronger than before. In the outburst of sorrow caused by that ' Parricide ' (as it was sometimes called by the lower classes), many village communes determined, as a sign of their grief, to close the drinking-shops. In three places in the government of Pskov a resolution to this effect was signed by 227 heads of families, and it was decided to close compulsorily all the public-houses, which have been taking 50,000 roubles a year from the population. In the government of Penza, where the governor has energetically striven to close these shops, the villagers declared in favour of abolishing them for ever. Three villages in the government of Vilna, moved chiefly by religious motives. did the same thing. General sympathy greeted that movement, for, as a rule, the smaller the number of drinking-shops the greater is the prosperity of the place. According to an interesting monograph of MM. Bektieff and Khvostoff on the economical position of Yeletz in the Ural, an examination of nineteen communes showed that, as a rule, the number of ruined homes corresponded to the number of public-houses in a commune. They mentioned two places as examples. The village of Jarnova possessed 203 homesteads and three public-houses. The soil was good ; the holdings

of each peasant averaged 4½ desatins per head. They
paid two roubles per desatin. After the public-houses
had been open for some time, 13 per cent. of these
peasants were entirely ruined, 25 others had no horse,
and 53 had not even a cow. As the possession of at
least one horse and one cow is the minimum of
prosperity, 78, or 38 per cent., of the peasants of Jarnova
had not even attained that minimum. Contrasted with
this sad spectacle of poverty was the state of the smaller
village of Petrovskoyè, which fortunately was without
any public-houses. Of its 55 homesteads only one was
entirely ruined, and only 4 were without a cow. Yet
the peasants only owned 2 desatins of land, and paid
for it 3 roubles 73 kopecs. Thus, although they had
to pay more per desatin, and only owned half the quantity
of land held by those of Jarnovo, only 7 per cent. are
below the minimum of prosperity, as against 38 per
cent. in Jarnovo.

'The same contrast, MM. Bektiff and Khvostoff report
is to be found in all the other villages they examined.
The wine-shops (*kabaki*) are now regarded as the village
cancers, and some of my friends in Russia would be
enthusiastic supporters of the United Kingdom Alliance.
Mr. Katkoff's *Moscow Gazette* publishes almost daily long
columns in favour of very drastic measures against too
great facilities for the sale of wine. Mr. Akaskoff's *Russ*
is just as emphatic on the subject. But it is only natural
for such enlightened and cultivated patriots as those two
to take such a course. Let me mention two others, who
although they have risen from the lower classes, may
nevertheless play an energetic part in the direction of
this question. I mean Mr. Tichomiroff and his uncle
Mr. Labsine. At present they are at the head of their
large manufactory at Bogorodsk, near Moscow. They
employ a great number of workmen, but they will never
engage a single man who is not a total abstainer. Extra
tea is willingly provided, and the wages are rather higher
than usual, but still the results economically and morally

are most excellent. Both Tichomiroff and Labsine are men of very deep religious feeling, devoted to their country, genuine enthusiasts. The former has taken an active part in the Commission of Experts, and his speech impressed his audience with his simple and fervid eloquence. His invectives against the drink-shops were exceedingly vigorous, and really made him a very valuable ally in the temperance reformation.

.   .   .   .

'It was very fortunate that neither the Government nor the experts shared the extraordinary theories held about the usefulness of drink-shops; on the contrary, the prevailing opinion is positively opposed to them. The initiative, as is generally the case in Russia, has been taken by the Government. One of the first acts of the new reign was the appointment of a committee at the Ministry of Finance to decide what steps should be taken to prevent the abuse of spirituous liquors.

'This committee, after eleven sittings in August and September, drew up a scheme of temperance reform which, in accordance with the excellent rule adopted by the Emperor, was submitted to a special commission of experts, selected from the Zemstvos of the empire for their special acquaintance with the subject to be discussed.

'There were thirty-two members of this commission, to whom two were subsequently added by vote of the commission under the title of special experts. The session of this temperance reform parliament was opened by General Ignatieff on the 24th of September at the Ministry of Finance. In his address, after explaining the desire of the Government that the representatives of the Zemstvos should be consulted before any legislation was undertaken, he referred to the question of intemperance as follows: "The sale of spirits in Russia, under the existing conditions, tended rather to the abuse of liquor and to the ruin of the people than to the satisfaction of any of the needs of the latter. The Government is

P

resolved to take efficacious measures to put an end to this
sad state of things, and it hopes that you will aid it in
discovering the method of doing this without injuring the
revenue."

.   .   .   .

'The first resolution of our experts, which was carried
with only five dissentients, was in favour of giving to
the communes the right to open communal public-houses.

.   .   .   .

'The second point decided by the commission was in
favour of a reduction of the number of public-houses.
It was resolved that the Zemstvos and municipal councils
should have the right to decide the number, the size,
and the type of public-houses in their locality ; the right
to issue licenses for commissions to be reserved to a
special licensing board, composed of justices of the peace,
members of the delegates of the Zemstvos, marshals
of the *noblesse ;* and the normal proportion of public-
houses to population to be 1 to 1000, which is equivalent
to closing about two-thirds of the existing places of sale.
The Zemstvos are to have the right of increasing or
decreasing this proportion by 25 per cent. They are also
to have the right to close them altogether, or open more
than the normal number with the assent of the Minister
of Finance and the approval of the provincial assembly.
'They decided in favour of confining the sale of spirits
in the rural communes to two descriptions of shops. The
first are those with the right of sale for drinking on the
premises, as you say in England, which answers to your
ordinary public-house. The license is only to be given
to them on condition that they also provide tea and
food for their customers.
'The second description of shops are those for the sale
of liquor in corked bottles for consumption at home.
Hotels and railway buffets are to be left as they are.
Restaurants where drink is sold are to be limited in
number by the municipal councils.

'Every three years lists of localities where public-houses can be established are to be drawn up by the Zemstvos, with the assistance of the excise officers. A proposal to permit the local commissary of the police to assist in drawing up these lists was defeated after a very animated dispute.

'Another important decision was that which interdicted any member of the Zemstvos or municipal councils to hold a license for the sale of drink, and the owners of the houses where drink is sold are not to be permitted to vote in the settlement of any questions relating to the drink-shops. They are only allowed to have a consultative voice—a very necessary stipulation.

'It was also decided to forbid the opening of any drink-shop within less than 40 to 100 sagénes (from 90 to 230 yards) from any church, school, or other public building in towns, or within less than 100 to 200 sagénes in villages.

'Communes and individual proprietors are to be allowed to forbid the opening of drink-shops upon their own land.

'One of the decisions at which the experts arrived, with only one dissentient, referred to the sale of drink by Jews. Well, we do not like the Jews, that is a fact; and the dislike is reciprocal. But the reason we do not like them is not because of their speculative monotheism, but because of their practical heathenism. To us they are what the relics of the Amorites and Canaanites were to the Hebrews in old times—a debased and demoralised element which is alien to our national life, and a source of indescribable evils to our people. It is not to the Jew as a rejector of Christianity that we object; it is to the Jew as a bitter enemy of Christian emancipation, the vampire of our rural communes, the tempter of our youth, and the centre of the demoralising, corrupting agencies which impair our civilisation. Ask anybody who has lived, if only for a day or two, near our custom-houses, and you will learn that all the smugglers, all the receivers of stolen goods, all the

keepers of brandy-shops, are the degenerate descendants
of the great Semitic race. If the Jews but obeyed the ten
commandments of their Lawgiver, there would be but
little objection to them in Russia. But as even Moses
found his Jews more than he could manage when his back
was turned, it is perhaps not surprising that Russians
have much difficulty in managing a people in whose ears
the thunders of Sinai have long since grown faint.

'The *Pall Mall Gazette* recently, in a fit of noble
indignation, delivered a very long lecture on the cruelties
of Jew-baiting in Russia. It might have had some
weight if the writer had not been as inaccurate as he was
prejudiced. For instance, Russians were solemnly
upbraided for confining the chosen people to " the most
ignoble occupations." No doubt. But considering the
number of Jewish journalists in Russia, the editor of the
Pall Mall Gazette does not seem to think much of the
dignity of his profession. But Jews are not only
journalists with us ; they also follow the equally "ignoble"
occupations of professors, teachers, authors, lawyers,
barristers, doctors, bankers, merchants, to say nothing of
those who occupy positions in the Government service !

.   .   .   .

'An intelligent diplomatist, who has lived a long time
in Russia, said to me the other day, when we were
discussing this question, " The forbearance of the Russians
is wonderful. No one can imagine how much they have
suffered at the hands of these Jews. It is strange that
these outbreaks have never occurred before." But it is
by no means only Russians who find it difficult to love
the Jews.

.   .   .   .

'There is one "ignoble occupation," however, to which
the Jews are very much devoted. The Jewish papers
declare that no fewer than one hundred thousand Jewish
families will be ruined if the Jews are not permitted to
keep open these infamous drink-shops which are the

curse of the Russians communes. How many hundred thousand honest Russian families, I wonder, have these Jewish brandy-sellers ruined?

'That our objection is solely to the anti-national Jews, not to Jews who become Russians in all but their origin, is proved by the decision of the commission in favour of allowing the Karaïte Jews, or "Karaïmes," as they are called, and call themselves, in Russia, to sell drink as freely as any other of their Russian fellow-subjects. It is only the Talmudist Jews who are forbidden that privilege.'*

This is considered a fair statement, of what is alleged against the Jews in extenuation of the bad feeling of the peasants towards them. With regard to the bad feeling of holders of landed property it appears that many of them, in borrowing money, find it convenient to do so from Jews, mortgaging estates in security for the loan; and if they do not repay the loan the mortgage is foreclosed; and Jews become the purchasers. Many do so; and the stranger may be ready to ask, And why not?

---

* A correspondent of the *Pall Mall Gazette*, who boasts an intimate knowledge of Hungary, says:—'Throughout the whole of Hungary hardly a public-house or village inn can be found which is not owned by a Jew. The Hungarian lower classes, like the English, are unfortunately addicted to drinking; and it is by skilfully taking advantage of this vice that the Jews make their fortunes, and at the same time raise such ill-feeling against themselves. This is how the matter works. A peasant enters a public-house in the evening intending to spend the few kreutzers he may have in his pocket on drink As soon as these are spent he will very likely get up to go—I have been a witness to this scene more than once myself—but this does not suit the landlord's purpose, who will say to him " Stay a little longer and I will chalk up what you drink down." The peasant—already, perhaps, a little excited—cannot resist the temptation, and before he has left that evening the commencement of a long score is already made. The next time he finds it so pleasant and simple to drink without paying that he allows his score still further to be increased. This goes on till the peasant is in debt for a considerable sum. Then the Jew turns round, his former civility changes into menaces. Finally he consents to allow the matter to stand over, on the peasant giving security on his land for principal and interest of the debt. A fresh score is run up, the interest is not paid, and at last the Jew seizes the peasant's land; for in Hungary, it must be remembered, every peasant owns a piece of land. In this manner all the peasant hold-ings are gradually but surely passing into the hands of the Jews. In the village of Cziffer, for instance, and several more places could be quoted, more than one-third of the lands formerly belonging to the peasants is now owned by the Jew landlords. Any one who knows the deep love the Hungarian peasant has for his land can readily imagine the strong feeling of hatred he will cherish towards the class who have robbed him of it by such means. And it is to this cause, more than any other, that the present disturb-ances against the Jews are to be traced.' The *Gazette's* comment upon this is that if this were the only ground for Jew-baiting the anti-Semites might more intelligently direct their energies to making tavern scores in Eastern Europe irrecoverable by law. This is the case in England at present.—J. C. B.

' The complaint of the patriotic advocate for the conservation and economic exploitation of forests again is that the Jewish proprietor does not choose to maintain the estates in the condidition in which he buys them; and does not choose to do anything to develope the agricultural capabilities of the land; nor does he choose to adopt the most advanced method of forest exploitation, but clears off the timber by felling *a blanc etoc*, and leaving the land to recover itself as best it may, while consequent droughts, and, it may be inundations, devastate the district and the lower lying lands. All this may be saddening; but with existing ideas in regard to the rights of property, who can say him nay?'

My purpose is to report facts. It does not come within the scope of my scheme to correct abuses in the lands upon which I report; but I may state here, as I have stated elsewhere, with a view to the discussion of what may, or might, be done to prevent the occurrence of like evils in other lands, that it is a principle accepted by many students of forest science, that the continued existence of forests in certain circumstances is so essential to the well-being of the nation inhabiting the land, that they should be considered national property, held beneficially in trust by the holders, and by the generation of which they are a part, for the nation in its entirety, of past, present, and future existence, each generation successively having the usufruct, but nothing more, being bound to hand them down to those who come after them in like good condition as they receive them, or with compensating advantages in one form or another for any destruction of them which may be deemed expedient; and that the exploitation of all forests should be subjectable to legislation with due regard to the rights of private proprietors.

CHAPTER VI.

Of the game found in Lithuania, Mr Anderson, already cited, gives the following account:—
'The noblemen and landed gentry of Russian Poland are very fond of hunting and shooting; but their mode of following these sports differs very much from that which prevails in England. For some time past, indeed, shooting has almost entirely ceased, in consequence of the public prohibition of the Government to carry or use a gun. To some favoured few, a licence to do so has been granted; the licence being sealed on the stock of the gun or rifle. But the commanders of the district towns generally advise the possessors of such a licence not to avail themselves of it; for the sound of fire-arms cannot fail to attract the Cossacks and Russian soldiery : and, as many of them are unable to read, the life of the poor sportsman, if he fell into their clutches, would not be worth five minutes' purchase. In consequence of this state of things, I only fired a gun upon one occasion, whilst I was in the country—under circumstances which I shall notice hereafter.

' The chief birds of game are the capercailzie, black cock, and wood hen—a bird very like the grouse, only smaller, and of a much lighter colour. It is called in German, *haselhuhn.* This bird lives in the woods, and is very seldom found, like our grouse, in the open. Of the common brown partridges—and this year was very favourable to them—we saw, frequently, large and numerous covies. The red-legged partridge is never found. The quail, woodcock, and snipe are very plentiful; and, on a summer's evening, the landrail may be heard in full "crake, crake." There are immense quantities of wild-fowl of all kinds. The

bustard also abounds in the country, and is considered a
great dainty; but its shyness makes it very difficult to
approach.   I was fortunate enough to catch a good view
of the first I ever saw, for we came suddenly upon him as
we were driving one day, about the middle of April.
During the summer, I remarked several flocks of them
at a distance.

' Besides the fox, the badger, and the hare, which this
country possesses in common with England, it has large
numbers of elks, buffaloes, bears, and wolves.   The elk is
rarely met with in the southern part; but, in a large forest
near Grodno, there are several.   During our stay at Wiercieliszki, part of the wood was on fire, and the flames
disturbed some of its wild inhabitants.   Word was brought
to Count Bisping that two elks had passed across his farm,
upon which he immediately mounted his horse and set off
in pursuit, with some big black hounds.   He failed to
overtake the elks, but plainly marked their track.   They
had gone through a piece of standing rye; and, in the wet
soil, he pointed out to me the same evening, the clear
impression of their large cloven feet.   The buffalo, or
bison, is not frequently seen.   I was told that there was a
herd of them, about a thousand head, some forty or fifty
miles from Grodno, and they are very strictly preserved.
They are much larger than the American or African
buffalo.   The law forbidding the slaughter of one of these
animals is as strict as that which prohibits the murder of
a man.

' The wolf is greatly on the increase, as the inhabitants
are denied the means, which they formerly possessed, of
killing them.   One day, as we were drawing near a small
cover with some greyhounds, I observed a great number
of magpies and carrion crows, which, on our approach,
flew around, marking their displeasure at our intrusion by
cries and croaks.   We brushed through the little wood :
and, at the lower end, saw, what I at first thought to be a
dog trotting away.   I galloped after him, when my companion also saw him, and cried, *wilka, wilka,* (wolf, wolf.)

Our dogs, having sighted him, quickly caught him up ; but were shy of making further acquaintance with him, until urged on by the cries of the huntsman, when they soon rolled him over. The wolf had tried at first to gallop off ; but, unfortunately for him, he had partaken too freely of his breakfast, and could not escape from his swift and strong pursuers. He was quickly despatched with the butt end of the huntsman's whip. We proceeded with our trophy homewards, and, upon reaching a hamlet belonging to the property of Wereiki, the peasants met us, expressing the greatest delight at the death of their enemy. They told the Count that hardly a night passed in which the wolves did not rob them of sheep ; and that, two nights before, they had made off with a cow. Two old wolves with four young ones had been lately seen by them ; and, no doubt, the one we had just killed was one of the litter. He appeared about eight months old ; more than three parts grown, and very strong ; the bone of his leg was very large. He was in colour a brown-grey and black, with light tan legs, and greyish eyes. He ran with his tail between his legs, just like a cowardly cur. In winter, wolves assemble in large numbers, and, being stimulated by hunger, are very formidable. But, as long as they are alone, and not much pinched for food, they are easily frightened. On one occasion, the wife of the Count's farm director was returning in her carriage from a friend's house, where she had been visiting, about five miles distant. One of her carriage horses was a mare, by the side of which (as the custom is in Poland) was running a little black foal. It was just becoming dark, when suddenly they were startled by seeing what they at first thought was a dog, running after the foal. But the coachman soon made him out to be a large wolf. He gave the reins to the lady ; and, jumping out of the carriage, picked up a goodly supply of stones. He then called the foal, which instantly ran up to him : for the foals, being always in the stables with the other horses, become tame as dogs. The coachman next turned round manfully upon his enemy, shouting at him, and pelting him with stones. The brute forthwith acknowledged the superiority of his assailant, and slunk away into the wood. The lady meanwhile was not a little alarmed by the reconnoitre, and right glad to reach home in safety. It was towards the end of August that this incident occurred, and wolves are rarely found to be so venturesome at this early period of the autumn. As the winter advances, hunger compels them to more daring deeds.

'Traps of all kinds are employed to catch the wolf in severe weather; the steel snap-trap, the pitfall, and the split tree—like that in which the old bear is represented as caught, in one of Kaulbach's illustrations of "Reineke Fuchs." The last, it is said, is the best snare. Sometimes, but very rarely, a fox is caught in it instead of the wolf; but the characteristic cunning of the fox generally prompts him to avoid it. Strychnine also is frequently used to destroy the wolf; and many become the victims of this poison. But the peasants more frequently injure their own property, by resorting to this process of destruction, for their dogs, being left to forage for themselves, are attracted by the poisoned bait, and die in consequence.

'The fox in Poland, as in Germany, is ingloriously murdered. He certainly has the pleasure in Poland of hearing the music of his pursuers, but his death by the gun is accomplished in a way which would be indignantly condemned by the English fox-hunter. As soon as the hunters are posted in the wood, the dog-keeper lets loose his pack of ten or twelve large clumsily-built hounds, of black or tan colour. They trot away without any order, and speak to every kind of game, hare, fox, or wolf. In brushing about the wood, they often start other game which, if it come in the way of the hunters, hardly ever fails of being shot; for these hunters are capital marksmen. The wild boar also is frequently started from his lurking-place on these occasions, and is always regarded as game of the first order by Polish sportsmen.

'There are two kinds of hare. The field or common

brown hare, and the wood hare, which is very like the Scotch hare, as it changes its colour with the season. I saw several of them quite white, when I first went to Poland; whereas, in summer, they are a brown-grey.

'The huntsman came home one evening with a large dog badger. It appeared that a hare, which he had been chasing with his dogs, took refuge in a small opening in a bank, which proved to be one of the entrances into the badger's hiding-place. As the hare ran in at one end, the startled badger sprang out at the other, almost into the jaws of the dogs; and was soon despatched by them and by the huntsman's whip. The poor hare also was afterwards pulled out of her hiding-place, brought home in a sack, and, after a few days, produced again to furnish sport (as it was called) for a brace of young greyhounds. But the confinement had so broken the spirits of poor puss, that she became, as might be expected, an easy prey to her pursuers.

'The common dogs of this country are a wretched mongrel race, and a most intolerable nuisance. They are to be seen in the house of every peasant, and crowding the streets of every village, yelping at the heels of every traveller, and flying out upon him, whether in carriage or on horseback or on foot, with great ferocity.

'The greyhounds are of a strong build, and far heavier than the thoroughbred animals seen at the coursing meetings in England. Many of them have long feather on their tails and legs; and these are much superior to the smoother sort, being quite as fleet, and endowed with higher courage and greater powers of endurance. The amusement of coursing is oftentimes greatly impeded by the quantity of rough and sharp stones with which the surface of the soil is covered. Indeed, I one day saw a poor greyhound so mutilated by one of these stones that it was found necessary to destroy him.

'The pointers are, as a class, very inferior dogs. One, indeed, was to be regarded as an exception—a coarse, heavy animal in appearance, but with a most sensitive nose, and excellently trained by his master, the Count's huntsman, who was a keen and sagacious lover of field sports.

' We took this dog one day upon an expedition which to me appeared very like a poaching enterprise. In walking over some marshes at Wiercieliszki, we had seen a great many snipe, some of which Count Bisping was resolved to have. Accordingly, he had a net made of very fine thread, about twenty yards in length and width. We began work about eight o'clock in the morning, the weather being very mild. The dog found the snipe well, standing very staunch. The two men who had the management of the net ran over the place at which he pointed, covering both the dog and game. He took it very calmly, and stayed until the game was captured. I never saw birds lie so close. They would not get up, even when the net was over them. Indeed, we lost several at first, thinking that it was not possible for any birds to be there, and not rise up in alarm ; and that the dog must have pointed false. On lifting up the net, away flew two or three birds from the very spot covered by the net, proving the dog's staunchness and the folly of our impatience. The whole arrangement, I must repeat, was very like poaching ; but in this country all is allowable. No net, no snipe, appears to be the rule.

' We were walking in the woods another day with this same dog, in search of blackcock, though we were not allowed to shoot them. The dog came to a good point; and, as we followed him up, away went a hen bird with five young ones, all fine birds. We marked down one of the young birds, and went after him ; and the dog was soon seen again pointing beautifully. The bird had crept into a small bush by which we were standing ; and, on its rising, flew directly in the face of one of our party, who hit at it almost involuntarily with his stick, and it fell into the dog's mouth. I confess, I felt ashamed at this mode of bagging game ; but I was told, that, at the present time, the capture of game, by any means, is accounted lawful.

' Permission was once granted to us, through the kindness of a Russian Director of the Government woods, to have some shooting. Count Bisping's huntsman, who, I have said, was a keen and experienced sportsman, had been obliged, on account of the insurrection, to give up his gun to this same Director; and it was with no ordinary satisfaction that he came to us one day, towards the end of April, with a message from the Director, saying, that, if the Count and I would like to have some woodcock shooting, we were to be at his house by five o'clock the next evening The Count was unwilling at first to embrace the offer, fearing lest he might thereby compromise the Director or himself; and that the noise of fire-arms in the forest might lead to collision with the Russian soldiers. But, on further consideration, feeling assured that the Director would not have sent such a message without ample authority, he agreed to go. Accordingly we started, at four o'clock, in an old post-cart without springs, and a pair of horses; and soon reached the comfortable house of the Director, who was by birth a German, and an intelligent and agreeable man. He offered us coffee and cigarettes; and showed us his private room, hung round with various trophies of his success in the chase : the antlered head of the elk, the smaller head of the roe, the tusks of the wild boar, the skins of the fox and bear. I also observed in a stand some very useful double-barrelled guns and rifles, in excellent order. His equipment was after the style of German sportsmen, who always carry a game-sack, like a railway travelling-bag; the powder-horn gracefully suspended round the shoulder by a green cord (such as we use in England for Venetian blinds) with large green tassels. The shot-flask is carried in the pouch. The gun has a beautifully-ornamented sling (generally worked in worsted by some fair hand), with which they carry it hung round the neck; and it is, in my opinion, very much in the way.

' After all preparations had been duly made, we mounted our waggons, having for our advanced guard a Cossack fully armed, and another equipped in like manner in our rear. The presence of these men of course removed any misgiving which might have been felt as to the authority under which we ventured forth ; and off we went at the usual galloping pace observed by travellers in this country. Our course lay through a large wood, with no very definite roadway ; and, as there had been lately some heavy rain, the ground was little better than a continuous bog. The horses sank up to their bellies three or four times, and how they ever came out again is still a mystery to me. And yet more wonderful does it appear that our rope tackle bore without breaking, the violent jerks and strains which it had to undergo. After about an hour spent in this hazardous journeying, we reached an open space in the middle of the forest, where we alighted and loaded our guns. Whilst we were thus engaged, a large hawk came and settled on a high tree close by. One of our party, a young man, whose eye was as quick and piercing as that of the hawk, speedily brought him down. At this moment I heard a curious noise, like hammering, which seemed gradually to come nearer ; and, upon asking what it was, learnt that it proceeded from the large wooden bells fixed on the necks of cattle which feed in the wood, and some of which I saw a few minutes afterwards. The sound of their bells is disagreeably monotonous.

'The woodcock begins to fly about half-past six o'clock, and flies for about an hour ; so we had not much time to lose. The huntsman soon posted us at our various stations ; and, during the few minutes we remained thus waiting, I heard distinctly the crane whistling, and the capercailzie crowing. But very soon a whirring, chattering sound announced the approach of the first woodcock. Its slow flight seemed to offer an easy shot ; but the dusky light balked our aim ; and the first three or four shots, on the part of the huntsman and myself, were failures. We were afterwards more successful ; and three birds fell to my share. I saw several others, but at too great a distance to reach.

'It was the beginning of the breeding season when we went upon this expedition ; and, according to our English notions, the pursuit of any game at such a time was unlawful ; but it is not so regarded in this country.

'We returned home to Wiercieliszki by a smoother and more agreeable road. The Director joined us at supper, and proved himself, by his amusing stories, to be not less welcome a companion than he had been a kind and zealous sportsman.

'The crane is a bird often to be seen in this country. They gather together in flocks, amounting to many hundreds ; delighting chiefly in marshy ground, over which they stalk gently with light and graceful step. neither experiencing, nor appearing to fear, any molestation.

'The stork also cannot fail, from its novelty, to attract the notice of the English traveller. The first stork's nest I ever saw was at Marienburg, on my way from Berlin to Konigsberg ; and, at that season of the year, it was, of course, untenanted. But the stork is held in great reverence by the people among whom it takes up its abode. Much pains are likewise taken to preserve the stork's nest, and encourage the parent birds to return to the spot which they have once selected for hatching and rearing their young. The storks are often very lazy in constructing their nests ; and the people consequently help them, by putting up an old harrow, or something of the kind, on the roof of a barn, or the top of some unused chimney, or the branches of a solitary tree. Over this they spread a foundation of hay or straw, and then pile upon it some twigs, or loose sticks, placed across each other, after a rough fashion, to about the height of two feet. This place the storks accept for their nest ; and, upon their arrival, will sit for hours, as if resting themselves after the fatigue of a long journey. They will then set about the task of putting the nest into something like order ; and, in about a fortnight, if the weather be warm, will begin the work of incubation. The parent birds

never both leave the nest at the same time; and the
male takes his turn on the eggs as well as the female.
They may be seen for hours, with their long bills projecting over the edge of the nest, patiently performing this
duty.　Nobody ever thinks of disturbing them, there or
elsewhere.　They may be seen sometimes, three or four
in number, striding quietly after the husbandman, as he
works away with his bullocks and plough, and securing
for themselves a meal from the worms, which the up-
turned furrows expose to their view.

'It is curious to observe the process by which the
storks feed their young.　Each parent bird goes away in
turn ; and, upon its return, stands, for a few seconds,
balancing itself upon the edge of the nest; then, throwing
back its head with a quick action, it ejects from its crop
into the nest some portion of worm or frog which it has
picked up ; and the young instantly seize upon the same
and devour it with avidity.　This action of throwing the
head back and ejecting the treasured food is repeated,
until all the contents of the bird's crop are exhausted.
About the middle of September, the storks assemble in
large flocks, like the swallows in England, and prepare for
their migratory flight to warmer latitudes.

'I have elsewhere described the manner in which the
Poles sometimes spear fish at night time, by the aid of
fire-light, which they kindle in a grate fixed at the head
of their boat, and need not therefore do more in this place
than allude to it as one of the modes to which they resort
for capturing fish.

'In England I used to be very fond of fly-fishing, and
had brought with me a nice light rod, in the hope of
meeting, in the streams of this country, with trout and
grayling as abundant as those which I had found in some
of the affluents of the Rhine.　But the hope was not to
be fulfilled.　As for trout, I was informed that it is found
only in one stream in Lithuania.　I sometimes caught a
few fish like grayling, but smaller and coarser, and of a
muddy taste.　Indeed, the constant muddiness of almost

all the streams in this country is alike destructive of the
delicacy of the fish, and of the pursuit of the fisherman.

'The pike is the most common fish to be found here,
and grows to a very large size in the extensive lakes
which abound in the country. The chief revenue, in fact,
of some properties, arises from their fisheries, and from
the number of their wild fowl. Some of the lakes are
four or five miles in circumference, and surrounded in
some instances with hundreds of acres, upon which
neither man nor beast ever ventures to set foot. The
proprietor has to pay as large a tax for this area of marsh
and water as for his arable and pasture land, and not
without reason; for the great demand for fish, in a
country peopled for the most part by Roman Catholics,
must make this kind of property very valuable.

'The carp is also found in these lakes, frequently of
large size. I saw one brought to table, which weighed
upwards of eighteen pounds; and, being stewed with
Hungarian wine-sauce, was very palatable.

'The tench is of a more delicate and agreeable flavour
than any fish in this country. I have frequently seen
them weighing from two to four pounds; and one even
reached the weight of eight pounds.

'It is no unusual thing to meet with pike ranging from
twenty to twenty-five pounds' weight, and sometimes
more. The largest are generally captured in nets; but
the quantity of weeds, with which the lakes and streams
are covered in summer, makes net-fishing very difficult, if
not impracticable. The fish are caught therefore in the
greatest numbers as soon as the weeds die away, and are
preserved in stews or in ice-houses.

'For my own part, net-fishing did not hold out much
attraction. I therefore persuaded my friend, the ingenious
carpenter at Wereiki, to fashion a dozen trimmers for me;
and, having prepared some twine and hooks which I had
bought at Konigsberg, I frequently made successful use of
them. I caught also, in a large lake which I visited
at Massalani, by trolling, many pike, which varied

Q

from one to eight pounds in weight. Upon one occasion, in June, I had sailed slowly, by the aid of a gentle breeze, from one side of the lake to the other, trolling as I went along; and had just started upon my return, when the bait was voraciously seized by a monster of a fish, that came right up to the top of the water, and darted off instantly with it. I let him have the line as quickly as possible. But the wind sprang up at this moment, and carried my boat off in one direction; whilst the fish was rushing madly away in another. I only had forty yards of line upon the reel; and, as soon as it had all run out, it of course snapped, and away went the only trolling gorge-hook that I possessed, with twenty yards of line.

' Thus abruptly and disastrously ended, as I thought, all my trolling speculations. But, upon returning to Massalani, a fortnight afterwards, the boatman brought me my hook and line, extracted from the fish, which be had found lying dead upon the water, a few days after my former visit. The fish, he told me, weighed more than twenty pounds. Behold ! the old fable of the ring of Polycrates over again—stripped, indeed, of much of its miraculous character; for a sharp gorge-hook was much more likely to lead to the discovery of the fish that swallowed it, than a precious jewel ; and the narrow circuit of a small inland lake a far more favourable place for such discovery than the waves of the open sea that washed the shores of Samos. Nevertheless, the incident could hardly fail to remind me, as it did, of the story, which has been told with such inimitable simplicity by Herodotus, and sung in inmortal verse by Schiller.'

PART III.

——:o:——

THE UKRAINE OR LITTLE RUSSIA.

To the district embracing the governments of Tchernigov and Kiev, with those of Poltava and Kharkov, are given the designations Little Russia, and the Ukraine. The name Ukraine is said to signify *frontier*, a name not inappropriate to the country, which lies on the borders of European Turkey, Poland, Russia, and Little Tartary. It was acquired by Russia at different periods from Poland, a large portion having been seized by Catherine II. in 1793.

Of the Ukraine M. Marny writes:—'In the Ukraine the black earth, called by the Russians *stepnoi-ezernozem*, which constitutes the soil of a part of South Russia, gives rise to forests of a special nature, of which the principal constituents are oaks, limes, and elms. These trees grow with uncommon vigour, and are associated with an immense number of large pear trees of a magnificent aspect. Nevertheless, this beautiful forest mantle is desolated under the pernicious action of drought, which causes the destruction of thousands of trees, more especially of hazels, ashes, and elms; only the spruce with deep roots escape its devastating influence.'

In Little Russia, and in the old Polish country to the south and the west, we find the villages of the north built of timber logs, and laid out as graphically portrayed in the quotation I have given from Hepworth Dixon's *Free Russia*, in *Forest Lands and Forestry of Northern Russia* (pp. 53-55), in illustration of forest scenes in the government of Archangel, have given place to villages of a different

character. We find the difference thus indicated
by that writer:—'Instead of the grimy logs, you
have a predominant mixture of green and white;
instead of the former blocks you have scattered cottages
in the midst of trees. The cabins are built of earth and
reeds; the roofs thatched with straw, and the walls of the
homestead are washed with lime. A fence of mats and
thorns runs round the group. If every house appears to
be small, it stands in a common yard, and one of its own.
The village has no streets. Two, and only two, openings
appear in the outer fence; one north, one south, and in
finding your way from one opening of the fence to another,
you pass through a mass of lanes between reeds and pines,
beset by savage dogs. Each new comer would seem to
have pitched his tent where he pleased, taking care to
cover his hut and yard by the common yard.

'A village built without a plan, in which every house is
surrounded by a garden, covers an immense extent of
ground. Some of the Kozack villages are as widely
spread as towns. Of course there is a church, with its
glow of colour, and of poetic charm.

'From Kiev, on the Dnieper, to Kalatch, on the Don,
you find the villages all of this type. The points of
difference [between these and those further to the north]
lie in the house and in the garden; and must spring from
difference of education, if not of race. The Great
Russians are of a timid, soft, and fluent type. They like
to huddle in a crowd, to club their means, to live under a
common roof, and stand or fall by the family tree. The
Little Russ [or inhabitants of Little Russia], are of a
quick, adventurous, and hardy type, who like to stand
apart, each for himself, with scope and range enough for
the play of all his powers. A Great Russian carries his
bride to his father's shed; a Little Russian carries her to
a cabin of his own.'

Conterminous with the government of Voronetz, on the
south-west, is the government of Kharkov, in which the

area of forests is 620,000 desatins, of which 308,883 desatins belong to the crown, equivalent to 12·9 desatins of forests, or 6·5 desatins of crown forests, to the square verst; and to 0·42 desatins of forests, or 0·2 desatins of crown forests, to each inhabitant. The annual fellings in the crown forests yield 37·8 cubic feet, and the revenue is 148·8 kopecs per desatin.

The government of' Poltava lies immediately to the west of the government of Kharkov. In this government the area of forests is 310,000 desatins, of which 39,142 desatins belong to the crown, equivalent to 7 desatins of forests, or 0·8 desatins of crown forests, to the square verst; and 0·2 desatins of forests, or 0·02 desatins of crown forests, to each inhabitant. The annual fellings in the crown forests yield 105·1 cubic feet, and the revenue is 228·2 kopecs per desatin.

In the government of Tchernigov, which lies to the north of Poltava, there are 928,000 desatins of forests, of which 261,697 desatins belong to the crown, equivalent to 20·1 desatins of forests, or 5·6 desatins of crown forests per square verst; and to 0·6 of forests, or 0·2 desatins of crown forest per inhabitant. The annual fellings in the crown forests yield 21 cubic feet, and the revenue is 30·1 kopecs per desatin.

In the government of Kiev, which lies to the west of the government of Poltava, the area of forests is 1,555,000 desatins, of which 272,739 desatins belong to the crown, equivalent to 26·2 desatins of forests, and 6·1 desatins of crown forests, per square verst; and to 0·6 desatins of forests, or 0·1 desatins of crown forests, per inhabitant. The annual fellings in the crown forests yield 21·2 cubic feet, and the revenue is 50·8 kopecs per desatin.

In the government of Tchernigov is the town of Baturin, a beautifully situated and well-built town on the banks of the Segma, named after its founder, Etienne Batori, King of Poland. It was formerly the residence of the Hetmen of the Ukraine. In 1703 it was the

rendezvous of the famous rebel Mazeppa, by whose over-
throw it suffered greatly, until, with the surrounding
villages, containing, with the town, a population of 20,000
souls, it was given by the Empress Elizabeth to the last
pet ward of the Ukraine, Count Cyril Razumoffsky, who
rebuilt it, and in whose family it still continues.

Of Kiev Mr Dixon writes :—' The first towns of Russia
are Kiev and Novgorod the Great ; her capitals and holy
places long before she built herself a Kremlin on the
Moskva, and a Winter Palace on the Neva. Kiev and
Novgorod are still her pious and poetic cities: one the
tower of her religious faith, the other, of her Imperial
power. From Vichgorod at Kiev springs the dome which
celebrates her conversion to the Church of Christ. In the
Kremlin stands the bronze group which typifies her
empire of a thousand years. Kiev, the oldest of Russian
sees, is not in Russia proper, and many historians treat it
as a Polish town. The people are Ruthenians, and for
hundreds of years the city belonged to the Polish crown.
The plain in front of it is the Ukraine *steppe;* the lands
of Hetman and Zaparogue; of stirring legends and
nations song. The manners are Polish, and the people
Poles. Yet here lies the cradle of that church which has
shaped into its own likeness every quality of Russian
political and domestic life.

'The city consists of three parts, of three several towns —
Podd, Vichgorod, Pechersk ; a business town, an imperial
town, and a sacred town. All these quarters are crowded
with offices, shops, and convents; yet Podd is the
merchant quarter, Vichgorod the government quarter, and
Pechersk the pilgrim quarter. These towns overhang the
Dnieper, on a range of broken cliffs ; contain about 70,000
souls ; and hold, in the several places of interment, all
that was mortal of the Pagan Duke, who became her
foremost saint.

'Kiev is a city of legends and events ; the preaching of
St. Andrew, the piety of St. Olga, the conversion of St.

Vladimer; the Mongolian assault, the Polish conquest, the recovery by Peter the Great. The provinces around Kiev resemble it, and rival it in historic fame. The country of Mazeppa and Gonta, the Ukraine teems with story, tales of the raid, the flight, the night attack, the revolted town. Every village has its legend, every town its epic, of love and war. The land is aglow with personal life. Yon chapel marks the spot where a Grand Duke was killed, this mound is the tomb of a Tartar horde, that field is the site of a battle with the Poles. The men are brighter and livelier, the houses are better built, and the fields are better trimmed, than in the north and east. The music is quicker, the brandy is stronger, the beer is warmer, the hatred is keener, than you find elsewhere. These provinces are Gogol's country, and the scenery is that of his story called " Dead Souls."

' Like all the southern cities Kiev fell into the power of Batu Khan, the Mongol chief, and groaned for ages under the yoke of Asiatic begs. These begs were idol worshippers, and under their savage and idolatrous rule the children of Vladimer had to pass through heavy trials; but Kiev can boast that in the worst of times she kept in her humble churches, and her underground caves, the sacred embers of her faith alive.

' Below the tops of two high hills, three miles from that Vichgorod in which Vladimer built his harem, and raised the statue of his Pagan god, some Christian hermits,— Anton, Feodosie, and their fellows, dug for themselves in the loose red rock a series of corridors and caves, in which they lived and died, examples of lowly virtue and Christian life. The Russian word for cave is " peck," and the site of these caves was called Peckersk. Above the cells in which these hermits dwelt two convents gradually arose, and took the names of Anton and Feodosie, now become the patron saints of Kiev, and the reputed fathers of all men living in Russia a monastic life.

' A green dip between the old town, now trimmed and planted, parts the first convent--that of Anton—from the

city; a second dip divides the convent of Feodosie from
that of his fellow-saint. These convents, nobly planned,
and strongly built, take rank among the finest piles in
Eastern Europe. Domes and pinnacles of gold surmount
each edifice; and every wall is pictured with legends from
the lives of saints. The ground is holy; more than a
hundred hermits lie in the catacombs, and crowds of holy
men lie mouldering in every niche of the holy wall.
Mouldering! I crave their pardons. Holy men can
never rust and rot! The purity of the flesh in death
is evidence of the purity of the flesh in life; and
saints are just as incorruptible of body as of soul! In
Anton's convent you are shown the skull of St. Vladimer;
that is to say, a velvet pall in which his skull is said to
be wrapped and swathed. You are told that the flesh is
pure, the skin uncracked, and the odour sweet. A line
of dead bodies fills the underground passages and lanes—
each body in a niche of the rock; and all these martyrs
of the faith are said to be like Vladimer—also fresh and
sweet! . . . Fifty thousand pilgrims, chiefly Ruthenians, from the populous provinces of Podolia, Kiev, and
Volhinia, come in summer to these shrines.

'When Kiev received her freedom from the Tartar begs
she found herself by the chance of war a city of Polonia,
not of Muscovy; a member of the western, not of the
eastern, section of her race. Kiev had never been Russ,
as Moscow was Russ; a rude barbaric town, with crowds
of traders and rustics, ruled by a Tartarised court; and
now that her lot was cast with the more liberal and
enlightened west, she grew into a yet more Oriental
Prague. For many reigns she lay open to the arts of
Germany and France; and when she returned to Russia,
in the time of Peter the Great, she was not alone the
noblest jewel in his crown, but a point of union, nowhere
else to be found, for all the Sclavonic nations in the
world.

'As an inland city, Kiev has the finest site in Russia.
Standing on a range of bluffs, she overlooks a splendid

length of *steppe*, a broad and navigable stream. She is the port and capital of the Ukraine, and the Malo-Russians, whether settled on the Don, the Ural, or the Dnieper, look to her for orders of the day. She touches Poland with her right hand, Russia with her left; she flanks Galicia and Moldavia, and keeps her front towards the Bulgarians, the Montenegrins, and the Serbs. In her races and religon she is much in little; an epitome of all the Sclavonic tribes. One-third of her population is Muscovite, one-third Russian, and one-third Polack; while in faith she is Orthodox, Roman Catholic, and United Greek. If any city in Europe offers herself to Pansclavonic dreamers as their natural capital, it is Kiev.'

In M. Polytaief's account of forests on the Dnieper, embodied in a preceding chapter, is given [ante pp. 187-190] information in regard to the forests of the government.

One of my correspondents, writing to me some years since, stated that shortly before he had had a conversation with a German land steward, from the neighbourhood of Kiev, who had the charge of a number of estates in that locality, and that in reference to the injury and loss to a country resulting from the clearing away of forests, he said the climate in that district had changed much for the worse, and that he attributed the recurring famines from which the people suffered, and the bad harvests of which they complained, in a great measure to aridity of soil and climate, consequent on the extensive cutting down of the woods. Of the Russian landholders, he said they were so improvident, and so reckless of the morrow, that it was difficult to induce them to plant trees which will not attain their full growth till some 170 years hence. And he alleged that there was no hope of the evil being counteracted, excepting in so far as necessary measures might be undertaken and carried out by the government.

Pinkerton epitomising a journey made by him from

Odessa to Orel, writes:—'The appearance of the country from Odessa to Nicholaief is flat and woodless. Near Elizabethgrad several ravines produce some little variety on the surface, but few trees of any kind are to be seen; and this bald uninteresting scene continues as far as Kiev, though with rather more undulations on the surface of the country as we approach the banks of the Dnieper. Those about Kiev may be styled hills; and are formed by the torrents that flow into the above-mentioned river, creating in their course a number of ravines. After crossing the Dnieper at Kiev, the country continues level the whole way to Orel, except here and there a few gently rising slopes; but there is extremely little wood, though the scenery is not quite so barren as that south of the Dnieper. Here there is little variety of prospect, therefore, to amuse the traveller excepting the common appearances of a regular Russian winter, the blue expanse of heaven above, with the woodless snow-clad earth beneath, and the clear pale powerless lunar-like rays of the sun shed over it; spotted with hamlets and villages, often at many miles distance from each other, and nearly concealed from view by winters revengeful robe; now and then a chain of sledges traverses the scene, and at the twilight seems to be moving in the air, so singularly does the united line of snow and sky deceive the eye of the beholder; or perhaps the sound of a fellow-travellers bell is heard tinkling, as he draws near. This precaution is adopted to prevent accidents in the night, as the movement of the sledge upon the snow, like that of a ship upon the sea, gives no indication of its approach.'

Such is the monotonous scenery which presents itself to the traveller in Russia everywhere during winter. But the intensity of the cold, and its invigorating effects upon animal life, seems to make every living creature to move with greater celerity; and in general I have observed the Russians give greater demonstrations of joy at the commencement of winter than at the opening of spring.

PART IV.

——:o:——

——:o:——

In M. Polytaief's account of forests on the Dnieper, it is stated that from the government of Volhinia the forest produce is sent to Riga and Prussia, with which there is river communication.

Lithuania is situated on the water shed between the valleys of the Black Sea and the Baltic ; the Ukraine is situated entirely on the former ; Poland almost entirely on the latter. While much of the timber yielded by these countries is conveyed by the Dnieper to the south of Russia, and much of what is produced by Lithuania and Poland is conveyed to Central Europe by rail, much also is exported from Dantzig, Konigsberg, Memel, and Riga. From Poland much of the timber is conveyed by the Vistula to Dantzig.

Of later exports of timber from this port I have no information. But from a report made to the Commissioner of Crown Lands in Quebec by Mr William Quinn, Supervisor of Cutters, compiled from notes collected by him on a mission to Europe, we learn that—

The supply of wood to Dantzig from Poland in 1860 was—

162,769 Pieces of full-sized square fir timber, being .... 45,943 more than in 1859.
101,737   ,,   ,,   small-sized square fir timber, ,, ... 37,899 ,, ,,
 15,081   ,,   Whitewood, square, ,, ... 7,605 ,, ,,
205,800   ,,   Roundwood, fir, ,, ... 70,320 ,, ,,
243,218 Fir sleeper logs, ,, ... 7,651 ,, ,,
 21,982 Oak planks, 1st brack, ,, ... 5,802 ,, ,,
 21,702   ,,   2nd   ,,   ,, ... 5,640 ,, ,,
 40,351 Pieces oak timber, planking logs and crooks, being .... 11,758 ,, ,,
 15,724 Shocks of oak staves, ,, ... 986 ,, ,,

And the exportation of wood from Dantzig in 1860 was—

229,190 Pieces of full-sized square fir timber, being .... 51,060 more than in 1859.
 57,127   ,,   small-sized   ,,   ,, ... 31,052 ,, ,,
877,392 Sleepers and sleeper logs, ,, ... 372,745 ,, ,,
326,987 Fir deck deals, deals and deal ends, ,, ... 37,833 ,, ,,
  2,066 Masts, spars, bowsprits, &c., ,, ... 9,305 less
  4,783 Fathoms of lathwood, ,, ... 844 more
 29,346 Oak planks, 1st brack, ,, ... 11,097 ,, ,,
 29,741   ,,   2d   ,,   ,, ... 6,141 ,, ,,
 96,083 Unbracked oak planks and plank ends, ... 41,847 less
 36,755 Pieces of oak timber, planking logs and crooks, being .... 2,753 ,, ,,
 14,091 Shocks of oak staves, ,, ... 1,740 more ,,

The stock of wood goods on the 31st December 1860 was—

 87,719 Pieces of full-sized square fir timber, being .... 18,816 more than on 31st Dec. 1859.
 74,408   ,,   small-sized square fir timber, ,, ... 19,451 ,, ,,
 11,951   ,,   whitewood, square timber, ,, ... 3,014 ,, ,,
154,113   ,,   roundwood, fir, ,, ... 34,085 ,, ,,
 13,757 Oak planks, 1st brack, ... 6,185 less
  9,438   ,,   2d brack, ,, ... 7,079 ,, ,,
 54,836 Pieces oak timber, planking logs and crooks, being .... 3,030 ,, ,,
  9,551 Shocks of oak staves, ,, ... 3,647 ,, ,,

THE EXPORTATION OF WOOD GOODS FROM DANTZIG TO THE DIFFERENT COUNTRIES IN 1860.

	TO GREAT BRITAIN.			TO FRANCE.			TO OTHER COUNTRIES.		
		More than in 1859.	Less than in 1859.		More than in 1859.	Less than in 1859.		More than in 1859.	Less than in 1859.
Pcs. of full-sized sq. Fir Timber..	190345	34352	....	19137	6068	....	19699	10640	....
,, small-sized ,, ,,	36572	22270	....	7376	243	....	13179	8539	....
Sleepers and Sleeper Logs..	722752	224489	....	....	....	....	154640	148256	....
Fir D'k Deals, Deal Ends & Deals	72553	....	16838	45504	3707	....	212930	50965	....
Masts, Spars, &c...............	385	27	....	1496	....	9322	185	....	10
Fathoms of Lathwood............	4782	916	....	....	....	....	1	....	72
Oak Planks and Plank Ends......	56465	....	42413	88778	15902	....	9927	1902	....
Pcs. of Oak Timber, Crooks, &c...	12887	....	410	5337	....	....	18531	1019	....
Shocks of Oak Staves..........	9469	3878	....	3289	....	....	133	....	187

The number of ships lying in Danzig on the 31st December 1859, was 124

,, ,, built in 1860, ... 6

,, ,, arrived in the course of 1860, ... 2542

In all, ... 2672

Sailed thence in 1860, ... 2576

Lying there on the 31st December 1860, ... 96

In all, ... 2672

Sailed to ....1267 ships, of which 641 with timber, 607 with grain, 19 with other cargo, ballast.

Sailed to	Ships	With timber	With grain	With other cargo	Ballast
Great Britain,	1267	641	607	19	
,, Holland,	266	57	208	1	
,, Sweden and Norway,	198		185		13
,, Denmark,	191	65	122	3	1
,, Prussian Ports,	161	2	101	41	17
,, France,	103	86	16	1	
,, Hanover,	93	18	75		
,, Bremen,	89	51	37	1	
,, Belgium,	84	39	45		
,, Russia,	57	47		8	2
,, Oldenburg,	27	25	2		
,, Spain,	13	13			
,, Hamburg,	5		3	2	
,, Mecklenburgh,	4	4			
,, Lubeckia,	4		4		
,, Italy,	1	1			
,, Africa,	1	1			
,, America,	1	1			
	2565	1051	1405	76	33

The following were the prices at Dantzig in the spring of 1861—

Square Red Fir Timber.

	Per load.
Best middling, 25 feet average length, ...	55s. 0d.
Good „ 26 „ „	47s. 0d.
Common „ 27 „ „	42s. 0d.

The usual dimensions are 15 feet and upwards, averaging as above, by 11-11ths to 18-18ths inches square. Shorter average lengths might be supplied at a reduction in price, whereas greater lengths are scarce and considerably dearer.

Small-sized Square Red Fir Timber.

9-9ths to 10-10ths inches square, 28 feet average length.

Best middling, ...	45s. 0d.
Second „	36s. 0d.

Whitewood Square Timber.

11-11ths to 11-16ths inches square, 32 feet average length, ...	28s. 0d.

Sleeper Logs, Red Fir.

9-9ths inches square, 8 11-12ths feet long, ...	25s. 0d.
10-10ths „ 8 11-12ths	31s. 0d.

Round logs, 10 inch diameter, 8 11-12ths feet ong, cost 2s, 2d. per piece. Prices of other dimensions of sleeper logs and sleepers vary from .... 30s to 34s. 0d.

Deck Deals, Deals and Deal Ends, Red Fir.

	Per 40 run. ft. / Per 720 running feet.
Deck deals, crown, 25 to 50 ft, av. 33 ft, 11 to 12 in. wide, 3 in. thick, ...	20s. 0d.
Crown, brack, „ „ „ „	13s. 0d.
Deals, „ 12 to 24 „ 18 „ „	10s. 0d.
Crown, brack, „ „ 8 „ „	£13 0s. 0d.
Deal ends, crown, 6 to 11 „ „ „	7 10s. 0d.
„ „ „	10 0s. 0d.
Crown, brack, „ „ „	6 5s. 0d.

All other thicknesses, from 2 to 6 inches, are paid in proportion to cubical contents.

PRICES AT DANTZIG IN THE SPRING OF 1861—Continued.

Deals, crown, 1¼ in. thick, 6 to 30 feet, aver. 17 feet, 10 to 12 in. wide,	17s. 0d. } Per 120 running feet.
" 1 " " " " 18 " 19 " "	11s. 0d.
Crown brack, 1½ " " " " " "	10s. 0d.
" 1 " " " 18 " 19 " "	7s. 0d.
Masts, Red Fir.	
13 to 15 inches diameter, 45 to 65 feet,	2s. to 3s. 0d. } Per run. foot
16 to 20 " 50 to 70 "	3s. to 7s. 0d.
Lathwood, crown, 8 feet,	£8 0s. 0d. } Per fath.
" 4 "	3 0s. 0d.

The price of the 7 feet in proportion to 8 feet, and that of 6, 5, 4½, 3½, and 3 feet, in proportion to the price of 4 feet.

	Per load.
Oak Timber, straight, 9 to 16 inches square, 18 feet average length,	£4 15s. 0d.
Oak Timber Ends, " " 6 to 11 feet in length,	3 10s. 0d.
Oak Crooks, " " 14 to 15 feet average length,	3 5s. 0d.
Oak Planks, 1st brack, 2½ to 7 inches thick and above,	8 5s. 0d.
2 " " and above,	7 0s. 0d.
2d " 2 to 7 " "	5 5s. 0d.
Oak Planking Logs (Plancons), hewn, 27 feet average length, 10 to 15 inches scantling, string measure,	3 15s. 0d.
Two sides sawn,	4 10s. 0d.

	Per mille of 1200 staves.
Oak Staves, Crown Vistula Pipe, 2¼ to 3, 5 to 6, 66 to 72 inches,	130 0s. 0d.
2 to 3, 4 to 5, "	95 0s. 0d.
Brandy, 2¼ to 3, 5 to 6, 54 to 60 "	95 0s. 0d.
2 to 3, 4 to 5, "	70 0s. 0d.
Hogshead, 2¼ to 3, 5 to 6, 42 to 46 "	70 0s. 0d.
2 to 3, 4 to 5, "	52 0s. 0d.
Barrel, 2¼ to 3, 5 to 6, 36 to 41 "	60 0s. 0d.
2 to 3, 4 to 5, "	45 0s. 0d.
Headings, 2¼ to 3, 5 to 6, 28 to 32 "	40 0s. 0d.
2 to 3, 4 to 5, "	28 0s. 0d.
2¼ to 3, 5 to 6, 18 to 27 "	36 0s. 0d.
2 to 3, 4 to 5, "	26 0s. 0d.

	per 60 tren'ls.
Trenails, Oak, 2 feet in length,	4s. 0d.
Fir, 4 "	7s. 0d.

Other lengths of oak and fir trenails in proportion to their lengths.

Mr Quinn had letters of introduction to Her Majesty's Consul General in Dantzig, and to the several mercantile houses engaged in the trade.   He reports—

'All those parties seemed earnestly inclined to afford ᴍe all the information possible relative to the trade of this port.   They are straightforward, open, and candid men, and did not appear to have anything to conceal of a general character with respect to the business.   They all complain that the standing timber is fast disappearing, that it is rising in price at each and every succeeding sale, and that the distance they have to haul is constantly increasing.   Mr Grade, of the firm of Messrs Albrecht & Co., said timber not requiring to be hauled more than 12 to 15 English miles is considered handy to the river.   To have to haul 6 to 8 German miles (30 to 40 English) is by no means unusual.   Afterwards it has to be driven a great distance by a tortuous, tedious, and expensive route.   A great proportion of the lumber brought to this market is made a long way to the south and south-east of Warsaw, and much of it is brought from Galicia, in Austrian Poland.   The general custom of selling the standing timber is as follows :—A certain limit or circuit is sold, which is supposed to contain a specified number of trees, suitable to be made into timber, for a slump sum or for so much per tree.   The number of trees is generally over-rated, but such is the competition among purchasers that they submit to this.   The purchaser is bound to take off the quantity within a given time, if to be found ; but in no case is any deduction made.   He is not allowed to take more than the number stipulated for, should they even be there, without paying additionally for them. Every tree which is cut down counts, whether rotten or otherwise.   I went with Mr Albrecht and looked over all the lumber in the river, down to the harbour.   There was but little remaining after the spring shipments, and none of the new timber had then arrived.   It was expected in a few days.   The timber is separated into three classes —1st, 2nd, 3rd.   Mr Albrecht told me that to

R

get any considerable quantity of first quality is very
difficult and expensive, and scarcely any of it is to be had
without having to be hauled 30 or 40 English miles. The
value of first quality redwood here at present is 55s. per
load, free on board ; 2nd, 45s. ; 3rd class about 41s. per
load. The freights just then were very low, not more
than 15s. per load to the east coast of England. Large
quantities of redwood are now being sawn up by the
different establishments here into deck plank for the
English and French Governments. The prices paid by
the French Government are for 1st quality 21s. sterling
for 40 feet long, 3 inches thick, and 9 inches broad ; and
two-thirds that amount for 2nd quality. There must not
be any pith in those planks, and they must show heartwood the whole length, of at least 7 inches wide. I find
that the production of last winter does not exceed that of
the previous year. A considerable quantity of redwood is
also being prepared here, intended for the defences at
Southampton, England. The pieces are all to be 35 feet
long, 12 inches square, and to show a certain amount of
heartwood on all sides. The price to be paid is 65s. per
load, free on board—a price with which the sellers seem
well satisfied. The timber purchased from the Prussian
Government in almost all cases is cut down and squared
at their expense. A portion of the timber is also got out
round the full length of the trees. It is then sold by
public auction—the square timber by the foot, the round
timber by the piece. The latter timber is brought down
without being squared, and part of it shipped as spars.
The remainder is sawn and manufactured into different
descriptions of scantling.'

In Britain we hear as much or more of Dantzic wheat
as of Dantzic timber. The wheat is not produced in
the vicinity of Dantzic, but is brought thither from a
great distance—some of it from the south of Russia, and
the transport barges make a demand upon the product of
the forests in Lithuania and Poland, which is by no

means inconsiderable.   In Fullarton's *Gazetteer* I find it stated :—

'Dantzic has long been regarded as one of the principal granaries of Europe.   It was at one time the only port from which shipments were made to foreign countries of the wheat sent down by the Vistula; and although of late years exportation has also taken place to a considerable extent from Elbing, Riga, and Memel, yet Dantzic still enjoys the greater portion of this trade, exporting occasionally 500,000 quarters of wheat in the year, besides flour, rye, barley, meal, oats, and pease. There are two modes of conveying wheat to Dantzic by the Vistula.   That which grows near the lower parts of the river—comprehending Polish Prussia, and part of the province of Plock, and of Masovia, in the kingdom of Poland—which is generally of an inferior quality, is conveyed in covered boats, with shifting-boards that protect the cargo from the rain but not from pilfering.   These vessels are long, and draw about 15 inches water, and bring about 150 quarters.   They are not, however, so well calculated for the upper parts of the river.   From Cracow, where the Vistula first becomes navigable, to below the junction of the Bug with that stream, the wheat is mostly conveyed to Dantzic in open flats.   These are constructed on the banks, in seasons of leisure, on spots far from the ordinary reach of the water, but which, when the rains of autumn, or the melted snow of the Carpathian mountains in the spring, fill and overflow the river, are easily floated.   Barges of this description are about 75 feet long, and 20 feet broad, with a depth of 2½ feet.   They are made of fir, rudely put together, and fastened with wooden trenails; the corners are dovetailed and secured with slight iron clamps, the only iron employed in the construction.   A large tree the length of the vessel runs along the bottom, to which the timbers are secured. This roughly cut keelson rises 9 or 10 inches from the floor, and hurdles are laid on it which extend to the sides. They are covered with mats made of rye straw, and serve the purpose of dunnage ; leaving below a space in which

the water that leaks through the sides and bottom is
received. The bulk is kept from the sides and ends of
the barge by a similar plan. The water which these
ill-constructed and imperfectly caulked vessels receive is
dipped out at the end and sides of the bulk of wheat.
Vessels of this description draw from 10 to 12 inches of
water, and yet they frequently get aground in descending
the river. The cargoes usually consist of from 180 to 200
quarters of wheat. The grain is thrown on the mats, piled
as high as the gunwale, and left uncovered, exposed to
all the inclemencies of the weather, and to the pilfering
of the crew. During the passage, the barge is carried
along by the force of the stream, oars being merely used
at the head and stern, to steer clear of the sand-banks,
which are numerous and shifting, and to direct the vessel
in passing under the several bridges. These vessels are
conducted by six or seven men. A small boat precedes with
a man in it, who is employed in sounding, in order to avoid
the shifting shoals. This mode of navigating is necessarily
very slow ; and during the progress of it—which lasts
several weeks, and even months—the rain, if any falls, soon
causes the wheat to grow, and the vessel assumes the
appearance of a floating meadow. The shooting of the
fibres soon forms a thick mat, and prevents the rain from
penetrating more than an inch or two. The main bulk is
protected by this kind of covering, and when that is thrown
aside, is found in tolerable condition. The vessels are
broken up at Dantzic and usually sell for about two-thirds
of their original cost. The men who conduct them return
on foot. When the cargo arrives at Dantzic or at Elbing,
all but the grown surface is thrown on the land, spread
abroad, exposed to the sun and air, and frequently turned
over till any slight moisture that it may have imbibed, is
dried. If a shower of rain falls, as well as during the
night, the heaps of wheat on the shore are thrown together,
in the form of the steep roof of a house, that the rain may
run off, and are covered with a linen cloth. It is thus
frequently a long time after the wheat has reached Dantzic

before it is fit to be placed in the warehouses. The warehouses are very well adapted for storing corn. They consist, generally, of 7 stories, 3 of which are in the roof. The floors are about 9 feet asunder. Each of them is divided by perpendicular partitions, the whole length, about 4 feet high, by which different parcels are kept distinct from each other. Thus the floors have two divisions, each of them capable of storing from 150 to 200 quarters, and leaving sufficient space for turning or screening it. There are abundance of windows in each floor, which are always thrown open in dry weather to ventilate the corn. It is usually turned over three times a-week. The men who perform the operation, throw it with their shovels as high as they can, and thus the grains are separated from each other, and exposed to the drying influence of the air. The whole of the corn warehouses now left—for many were burned during the siege of 1814—are capable of storing 500,000 quarters of wheat, supposing the parcels to be large enough to fill each of the two divisions of the floors, with a separate heap; but as, of late years, it has come down from Poland in smaller parcels than formerly, and of more various qualities, which must of necessity be kept distinct, the present stock of about 280,000 quarters is found to occupy nearly the whole of those warehouses which are in repair, or are advantageously situated for loading the ships. Ships are loaded by gangs of porters with great despatch, who will complete a cargo of 500 quarters in about three or four hours. [*Jacob's Report in* 1826.] In 1845, the exports of wheat were 34,106 lasts. In the beginning of last century rye was the grain chiefly exported from Dantzic. In the preceding century its annual average export of rye amounted to 95,000 lasts. The grain warehouses, and those for linen and hemp, are situated, as already noticed, upon an island, formed by the river Mottlau on one side, and another branch on the other. There are three bridges on each side of the island, which are drawn up at night, excepting the two at the end of the main street across the centre of the island, communicating between the Altstadt and the Vorstadt. To guard these warehouses, there are from 20 to 30 dogs let loose at 11 o'clock every night. To keep the dogs within these districts, and to protect passengers from harm, at the end of each street leading to the main one there are high gates ; no light is allowed, nor any person permitted to live on this island.

'Among the chief exports from Dantzic are masts, corkwood, hemp, flax, potash, honey, wax, tallow, iron, steel, copper, lead, saltpetre, tar, amber, skins, furs, wool, and salt; but corn and timber have always been, and still are, the staple exports. In 1788, the following estimate was made of its commerce :—About 60,000 lasts of corn from Poland were annually exported, amounting, at 18 ducats per last, to about 2,080,000 ducats. The wood, masts, potash, hemp, flax, hides, linens, skins, honey, tallow, wax, and tobacco, from Poland and the Ukraine, were valued at nearly the same sum ; making an annual capital of 6,000,000 thalers vested in commerce. The profit upon this sum amounted to about 1,200,000 thalers, from which, deducting 150,000 for taxes, custom-house dues, and interest of money, there remained 900,000 thalers as annual profit. In 1839, the total exports and imports were valued at 13,236,118 thalers, or about £2,200,000. The restrictive duties imposed by Russia have directed much of the produce of the Ukraine and Poland into other channels.'

From Konigsberg and Memel, Prussian ports east of Dantzig, the timber exports consist largely of the produce of Lithuania.

At Konigsberg Mr Quinn obtained much information from Mr Hertsel, the British Consul, who had previously resided some time at Memel, and was there engaged in the timber trade. Mr Quinn reports :—

' He seems to be well acquainted with the affairs of the country, as connected with the lumber trade. He informed me that about one-third of the forest lands which supply this place and Memel belong to the Russian

Government, and about two-thirds to Polish and Russian nobles, and that almost the whole of said supply comes off Russian territory, scarcely any off Prussian. There is scarcely any possibility of arriving at the cost of bringing it to market, the business being altogether in the hands of the Jews, who hitherto had an understanding with the proprietors that the serfs on the estates should be employed in making and bringing it forward, and, consequently, the exporters at these ports neither know or care about the cost of production, not being interested in the same.   What effect the emancipation of the serfs will have on the trade remains yet to be developed.   The great timber-producing districts are comprised within an area of about 27,000 English square miles—a great portion of which has been cut over and over again, besides, there is a population within this circuit of from 1,600,000 to 2,000,000, and it is considered one of the best agricultural provinces in the Russian empire.   From these facts I infer that there is a limit to the timber even in this province.   The standing timber is gradually becoming scarce and dear.   The distance to haul is increasing, and it is thought that the emancipation of the serfs will have the effect of changing the nature of the trade altogether. In the first place, it is to be expected that much more of the land will be brought under cultivation, and, in the next, the men will not continue to work for the same small pittance they have hitherto been in the habit of receiving.   In fact, they seem not to be inclined to work at all.   In proof of this assertion I can state that large numbers of Germans, from the province of Pomerania, at the time I was there, were moving to Russia to supply the labour heretofore performed by that class.   This change in the condition of the serfs must raise the price of labour, and a corresponding rise in the price of timber, or a diminution in the quantity, must necessarily be the result.   Mr Hertsel further informed me that the country is now undergoing such changes that it is hard to say what ultimate effect such changes may produce.   One thing, however, is certain, that, so far, the effect has been a large diminution in all the products of the country.'

Mr Quinn subsequently visited Memel with every provision for obtaining information upon which reliance might be placed. The corroborative information which he there obtained amounts, he says in his. report, to the following :--

'Memel is supplied with lumber from Russia and Poland by the river Nieme. The lumber has such a long distance to be driven that it only reaches the market in September and October, which market is at a place called Russ, about 30 English miles from Memel, situate on the bay which lies between that city and Konigsberg. At Russ the wood is purchased by Memel merchants, and brought down at their risk and expense in large rafts of 1500 to 2000 pieces of square and round timber. When the timber arrives at Memel it is assorted according to quality and the views of the owners. There is a government system of classification, but it is not compulsory. The different kinds of timber brought to this place are :—

Fir Redwood, square.	Oak, square.
„ „ round.	„ Wainscot Logs.
„ Whitewood, round.	„ Staves.

'And in about the following quantities :—

Square Redwood, 12 to 18 inches square, 20 to 60 feet long ; very few pieces of the latter length or size ; general average about 30 x 13,..................	150,000 to 200,000 pcs. an'ly.
Round do,.....................................	150,000 „
Whitewood, round,..........................	20,000 „
Square Oak,..................................	10,000 „
Wainscot Logs,...............................	4,000 „
Staves, about	15,000 shocks of 60 pcs. ea.

'The square redwood is classified as under :—

Crown Timber, value at that time,.........70s. stg. per load.			
Best Middling,	„	„	.........66s. „
2nd „	„	„	.........54s. „

'The round wood, both red and white, is cut here by steam and wind-mills into 3 x 11, 3 x 9, 4 x 11, 4 x 9, 2½ x 7, and boards 1 to 1½ inches thick by 8 to 11 inches in breadth, and all generally of long lengths. They are classified as follows—

Redwood, Crown, was then worth £12 per St. Petersburg std.
,, second quality, ,, 7 ,,
,, third ,, ,, 6 ,,
Whitewood, Crown, ,, 7 ,,
,, second quality, ,, 6 ,,
,, third ,, ,, 5 ,,

'The latter deals compete with our spruce in the English market; but according to my opinion they are not as good, or equal in any respect. I was told by the gentlemen in the trade that a few years ago this whitewood was attacked by an insect, which has almost killed every tree. I have seen a considerable quantity of this kind of lumber in the log, and found it all to be perforated to the heart by grab-worm.

'Square oak, 12 to 14 inches square, and 20 to 50 feet long, general average, not more than 35 feet cubic, and classified as under—

Crown, was worth at that time .... 100s. per load.
Second quality, ,, ... 90s. ,,

'Wainscot logs, in lengths of 9 feet and upwards. This timber is sawn from pretty large trees, must be free from heart, and shaped thus, $\overline{\underline{10 \; : \;}}$); and must be at least 10 inches deep from the curve to the corner of the large flat surface.

Crown.—The value at that time was 5s. 6d. per foot.
Second quality, ,, ,, 3s. 6d. ,,

'Staves are ofthe following dimensions—

Pipe, 6 feet long, 6 x 3 inches, ⎫
Brandy 5   ,,   ,,   ,, ⎪
Hogshead, 4 feet long, 6 x 3 inches, ⎬   All reduced to
Barrel,   3   ,,   ,,   ,, ⎪   6 x 6 x 3
Heading, long, 2¼ ft. long, 6 x 3 in.. ⎭
   ,,   short, 1½   ,,   ,,
And were then worth, Crown, £140 per 1,200 pieces.
   ,,   ,,   ,,   1st brack, 115   ,,
Three pieces long heading counts one.
Four   ,,   short   ,,   ,,

'The provinces which supply Memel are Kovno, Augustoo, Bialystock, Vitepsk, Minsk, Wilna, and Volbinia—this latter province furnishing or yielding two-thirds of the whole. As far as I can understand, the timber trade is gradually declining here, and although the lumber is becoming scarce, that is not the only reason given for the cause of the trade languishing. The late season at which the timber arrives necessitates the holding over large quantities during winter, which is embarrassing in a monetary point of view. Nevertheless, at that time in Memel, as well as in the other ports in the Baltic, the people were as busy as possible sawing and preparing timber for the French Government, and complained that they were not able to get it fast enough.

'The following were the prices of lumber, free on board, at Memel in March 1861—

Crown Fir Timber, 12 in. ⎱ 25 feet average, at... ....70s. 0d. per load.
     ,,   11 ,, ⎰   ....68s. 0d.   ,,
First Midlg, ,,   12 ,, ⎱ 26   ....64s. 0d.   ,,
     ,,   11 ,, ⎰   ....62s. 6d.   ,,
Second ,,   ,,   12 ,, ⎱ 27   ,, ...54s. 0d.   ,,
     ,,   11 ,, ⎰   ....52s. 6d.   ,,
Inferior ,,   ,,   12 ,,   25   ,, ...45s. 6d.   ,,
Oak Timber.—Crown, 100s. 0d., Second quality, 90s. 0d. per 50 run'g ft.
Wainscot Logs.—Crown, 5s. 6d.,..................... 3s. 6d. per run'g foot.
Deals, 3 x 10½-11ths and 3 x 9 inch. averaging 17 to 18 feet.
Red, Crown, £12 ; Seconds, £7 ; Thirds, £6 6s. 0d. ⎱ per 750 run'g. ft.
White, ,,   7   ,,   6 6s. ,,   5 5s. 0d. ⎰ 3 x 10½-11ths in.
Staves.—Crown Pipe, £150, 1st Brack, .. 125 0s. 0d. per 1,200 pieces.

Memel, a Prussian port, is situated not far from the

frontier, where it is conterminous with the Russian province of Courland.

At about the same distance along the coast, within this province is Liban, destined it may be to be ere long, through the increase and extension of railways in Russia, the one great Russian commercial port on the Baltic.

Meanwhile Riga,. at the head of the sheltered gulf to which it gives its name, and situated at the mouth of the river Dwina, in Polish Dzwina, is the great Russian port for the export of timber from the territory traversed by that river, chiefly Lithuania. This river must not be confounded with the Dwina in the northern government of Archangel.

One of my correspondents connected with the timber trade in Riga wrote to me in answer to enquiries I had addressed to him—' As to the forests of the Baltic provinces, strictly speaking they are nearly all cut down, at least the larger sized trees are, and those now obtained have in many cases to be transported a distance of some 20 or 30 versts (14 or 20 miles), on axle or sledge. In general the latter mode of conveyance is preferred, as it is the cheaper of the two. From this you will perceive that a good deal depends on the kind of winter we have, as good sledge roads are desirable for the transport of timber to the edge of the water, by which it is floated down in spring at the breaking up of the ice. This department of the trade, I may mention, is almost entirely in the hands of the Israelites.' He goes on to say—' To give you some idea of the extent of the timber trade in this place I may mention that there are annually exported about $2\frac{1}{2}$,000,000 of railway sleepers, and about 300,000 logs of square timber, which are hewn in the forests, all the chips and branches of which remain there to rot, and make manure for the next forest that grows in their place. There is no attempt made to clean out such refuse, in fact there is no fostering care whatever employed in such things, it is entirely left to nature. Anything can be made, such as squared timber, railway sleepers, &c., and it is a very rare thing indeed to replace any of the forests that have been so wrought out, by planting young trees in their stead ; nature is allowed to do that herself, and of course will accomplish it in a series of years. There are still a few forests in these provinces held by the crown, the first cost of which is so high that at present prices it almost precludes merchants from working them to advantage.

'The most of the timber exported from Riga at the present time comes from the provinces of Smolensk and Vitepsk, along the banks of the river Dwina and its tributaries, but owing to the great distances at which some of the woods are situated from water communication with the river there is a good deal of extra expense incurred for the transport, and this is increasing yearly, consequently the price of timber is getting much higher to work out this one problem. In addition to the above exports there is another large demand which must make a big hole in the woods yearly, viz. :—There are 1½,000,000 of fir logs, averaging 24 feet in length, rafted down from the above-mentioned provinces annually, and cut up into 3 inch planks and boards, and exported to England, France, Belgium, Holland, and Germany. Notwithstanding these great quantities, I am told the increasing demands can be met for a great many years to come, subject to an increase in price owing to the extra transport by water communication, which is indispensable to the trade.

'As to the titles of official reports or documents issued by the Government there is no·one here can enlighten me on the subject, all saying they never heard of any such things.

There are no wholesale joinery establishments here at present, but formerly there was one in connection with a large saw-mill which was burned down about 18 months ago ; and that part was not rebuilt, as it was considered by the proprietors to be a bad speculation.'

I was supplied with a copy of a report by Mr A.

Kauffman, which had appeared in the *Rijscomy Cbaesdy Laesochozgaev*, and of which the following is the substance. It is headed:— *To what extent is the Export Wood Trade from Riga in the future warranted by the Present Condition and Management of the Forests in the West Dwina Districts.*

'The writer of this report thinks it necessary in order properly to treat the question of the future supply of wood for export trade, not to confine himself strictly to the consideration of the present supply of the west Dwina district sent to the port of Riga, but to embrace the principal sources of supply on the Baltic and Black Sea, to which wood is conveyed to, and exported from, the ports of Memel, Dantzig, and Stettin—in short, to consider the whole export supply of the west and south-west country of the Dwina. In order to this, it is necessary at the same time to remember and to state the different branches of the home trade which derive their supply from these sources, these are—the Kherson port and the Crimea; the arsenals and waggon factories of St Petersburg and Riga; and railway materials for their own and the southern net-work of railways. And over and above this there has to be considered the wood that is used by the inhabitants of these districts for building purposes, together with the enormous quantity used as fuel.'

There is given a tabulated statement of the quantity and description of materials sent to the ports of Riga, Memel, and Dantzig, on the average of the past three years, and the overland exports for one season; and with this there is given the number of trees (arranged according to their respective ages) that were cut down in one winter to supply this quantity. The following is a summary of exports :—

(1.) TREES OF MATURE GROWTH.

	No. of trees.	Cubic feet.
Pine and fir wood from 140 to 250 years old,...	1,250,450	= 50,018,000
Oak ,, ,, 120 ,, 260 ,, and over.	373,600	= 30,548,000

(2.) TREES OF MIDDLE GROWTH.

	No. of Trees.	Cubic feet
Pine and fir wood, from 100 to 140 yrs. old & over	1,439,800	= 43,194,000
Oak, ,, ,, 120 ,, 160 ,, ,,	503,200	= 20,128,000

(3.) YOUNG TREES.

Pine and fir, from 40 to 100 years old and over,	1,650,000	= 24,750,000
Oak ,, 60 ,, 120 ,, ,,	700,000	= 11,900,000

(4.) UNDERGROUND.

Pine and fir, from 15 to 40 years old,..............	2,550,000	= 15,300,000
Oak, ,, 25 ,, 60 ,, ...	906,000	= 9,050,000
Total,	9,373,050	= 194,898,000

There are then given the details and description of the wood used for home trade, and the respective quantities sent to such places as St Petersburg, Odessa, Nicolaieff, &c., in one year, the summary of which is given as follows :—

HOME CONSUMPTION.

TREES OF MATURE AGE.

	No. of trees.	Cubic feet.
Pine and fir, from 140 to 170 years old,..........	579,000	= 20,265,000
Oak, ,, 160 ,, 300 ,, ·........	13,500	= 9,940,500

TREES OF MIDDLE GROWTH.

Pine and fir, from 100 to 140 years old,..........	559,000	= 16,810,000
Oak, ,, 120 ,, 160 ,, ...	58,000	= 2,352,000

YOUNG TREES.

Pine and fir, from 40 to 100 years old,....	677,000	= 10.155,000
Oak, ,, 60 ,, 120 ,.	400,000	= 6,800,000

UNDERGROWTH.

Pine and fir, from 15 to 40 years old,.............	1,056,000	= 6,396.000
Oak, ,, 25 ,, 26 ,, ...	1,000,000	= 10,000,000
Total,	4,353,300	= 82,718,800

Having given these statistics, the writer goes on to show how grave the consequences of the present system of indiscriminate and wholesale destruction of the forests must inevitably prove, and urges that immediate measures should be taken to prevent further devastation. He states it emphatically as his opinion that if the whole of the trees in these districts, of the first and second growths—which form the staple of the export trade—

were to be divided by the number exported in one year it would be seen that the export supply would not possibly be sustained for a period of ten years. Taking a retrospective view, he states that in the year 1859 the exports of wood amounted to the value of 4,500,000 roubles, in ten years from which time it had increased to 15,000,000 roubles, and it has gradually gone on increasing during the past five years—thanks to the abolition of the duty—till it had in the preceding year reached the enormous sum of 32,000,000 roubles. Calculations are then entered into to prove that if the present system of the management of forests be persisted in, the utter annihilation of all the forests of the west and south-west country of the Dwina within the period of fifty years is inevitable. The rest of the report is taken up in an endeavour to prove to the owners of these large estates, &c., that their interests will in no way suffer by Government supervision interfering with the management of their forests, but that this will rather be the means of adding to the value of their property by increasing the price of wood, &c.

In an article in *Le Marchand de Bois* on the lumber trade of Riga, it is stated :—'Of the ports of the Baltic provinces, Riga exports more lumber and timber than all the others, consequently a statement of the amount of the trade done at Riga will give some idea of the extent of the lumber trade of Russia in Europe. The details of the exports of Riga for a series of years, from 1871 to 1882, are taken from a recent official document issued from the bureau of commerce. The timber and lumber reaches Riga by the Dwina and its tributaries, and comes from the provinces of Livonia, Esthonia, Courland, Mohilev, and Minsk, in Lithuania. Notwithstanding the extensive devastation of the forests, carried on with no assurance of their renewal, the vast forests which formerly covered these provinces yet contain large reserves of timber of the best quality, but their continued depletion, with no provisions made for their future restoration, should give rise to serious apprehension. The transport of the wood from the forests to the rivers, which is made by log waggons during the snows and frosts of the winter season, is attended with increased difficulties and cost when the winters are usually rainy, which is quite common; and the transport by water is also difficult on account of the shallowness of the waters and the strength of the currents; and in addition to this it is peculiarly difficult to safely run the rafts which are constructed to suit the special navigation of the streams. The large timbers are bound together and form the raft proper. On these are then placed laths, staves, clapboard, &c. It is said the manufactured stuffs brought to market in this way are limited in quantity. This is to be regretted, as persons who are acquainted with the interior of Russia state that it would be much to the interest of the proprietors if they would prepare on their places in the interior the greater portion of the flooring, laths, deals, scantling, &c., as it is well known that large quantities of timber useful for these purposes are left to rot. It is also stated that the *debris* left by the loggers in the pine forests might be profitably distilled. But to bring about such necessary improvements and progress, the interior must have the benefit of better methods of communication with the markets than by the inefficient and tedious rafting system.

'The rafts are formed on the ice before it begins to break up, in order that advantage may be taken of the first rising of the waters. When this occurs the rafts usually arrive at Riga from eight to fifteen days thereafter.

'During the period above mentioned, from 1871 to 1882, there were exported from Riga 87,377,512 pieces of timber and lumber, or an average of 7,281,456 pieces per annum. The progress made in this trade since 1871 is shown by the fact that in that year only 4,542,155 pieces were exported; whereas, in 1882, 9,184,199 pieces were exported.

'The total value, in silver, of the wood exported in that time was 122,342,708 roubles; a yearly average of

10,195,225 roubles. The value of the exports of 1882 was placed at 12,911,072 dollars. A silver rouble is worth about two shillings at present.

'Of the leading articles exported during the twelve years, there were in pieces, 3,651,095 staves and clapboards; 30,382,566 sleepers; 49,491,155 posts and flooring. With the exception of 47,769 pieces shipped to Africa, the whole amount was cleared for European ports. England received 58·7 per cent. of the total exports; Holland, 14·1; Germany, 10·1; France, 8·6; Belgium, ·5; and Portugal, 2·2 per cent. The balance was taken by Denmark, Sweden, Norway, Spain, and Italy. During the year 1882 the average freight charges, per St. Petersburg standard, to French ports were as follows:—Dunkirk, Boulogne, and Dieppe 9 dols. to 9·50 dols.; to Havre, Honefleur, and Rome, 10·50 dols. to 11·50 dols.

'On the whole, the timber and timber trade of Riga for the past several years has made notable progress, and the movement in 1882 was especially favourable. The trade of France with this port, though not nearly so heavy as that with Norway, is yet of no mean importance, as the transactions for 1882 reached the sum of 835,000 dols., and it is susceptible of considerable expansion.'

Further to the east, on the Gulf of Finland, are the ports of Revel, Narva, and Cronstadt, from all of which there is an export of forest produce, but this is not to any great extent, if at all, derived from Lithuania. The following memorandum supplied by Mr Quinn, in regard to charges made at the time of his visit (1860), for the transport of wood from ports which have been named, and from Windau and Pillau may be useful for purposes of comparision:—

Cronstadt to London, deals..................	40s. 0d. per std.
,, to Exmouth Bight, deals, ... } or to Shoreham,........ }	47s. 6d. ,, 47s. 6d. ,,
Narva Bay to London, deals and timber,	57s. 6d. ,,
,, to East Coast, deals,............	55s. 0d. ,,

S

Narva Bay to Grimsby, deals,............ ⎫	52s. 6d.	per std.
Or square sleepers,......... ⎬	17s. 6d.	per load.
Or round ,, ... ⎭	19s. 6d.	,,
„ to West Hartlepool, square		
sleepers, ... ⎫	15s. 6d.	,,
Deals, ... ⎬	47s. 6d.	per std.
Option desired of Grimsby ⎭	17s. 6d.	per load.
Riga Town (Bolderaa) to London, square		
sleepers or fir timber,......	21s. 0d.	,,
„ to W. Hartlepool or Tyne Dock, ⎫	17s. 0d.	,,
Or square sleepers,............ ⎬	20s. 0d.	,,
„ (Belderan) to a dockyard in ⎫		
Thames, timber, ... ⎬	21s. 0d.	,,
And masts,... .... ⎭	24s. 0d.	,,
„ to Portsmouth, timber, ... ⎫	23s. 0d.	,,
Masts, ... ⎬	26s. 0d.	,,
Libau to London, timber, deals, or		
square sleepers, ...	19s. 0d.	,,
Windau to London, Hull, or Grimsby,		
timber or square sleepers,...	17s. 0d.	,,
Memel to London, staves, ...	£17 0s. 0d.	per mille
Or to Grimsby, ...	15 0s. 0d.	,,
Or to Liverpool,............	18 0s. 0d.	,,
„ „ or East Coast, timber		
and square sleepers...	17s. 0d.	per load.
„ Chester, timber,........ .... ⎫	20s. 0d.	,,
Or to Dublin, sq. sleepers... ⎬	20s. 0d.	,,
Round sleepers,............ ... ⎭	22s. 0d.	,,
„ Torquay, timber and deals,......	18s. 6d.	,,
„ English Channel, between Dover		
and Southampton,..............	18s. 6d.	,,
„ British Channel, timber or		
square sleepers, ...	19s. 0d. to 19s. 6d.	,,
„ Wexford, 200 loads timber,......	23s. 0d. to 24s. 0d.	,,
„ Table Bay, or Algoa Bay, deals,	£7 15s. 0d.	,,
Or Dantzig to Newport or Car-		
diff, sq. sleepers or tim.,⎱	19s. 0d. to 23s. 0d.	,,
Or round sleepers, at...... ...⎰	20s. 0d.	,,
Pillau to Combwich Pill, sq. sleepers... ⎫	21s. 0d.	,,
Or to Drogheda, ... ⎬	22s. 0d.	,,
Dantzig to London or East Coast, timber		
or square sleepers, ...	16s. 6d.	,,
„ Sunderland or a Coal Port, ⎫		
oak timber,.................. ⎬	18s. 0d.	,,
Or to Hull,.................... ⎭	20s. 0d.	,,
„ Shoreham, ... ⎫	18s. 0d.	,,
Or to Milford, timber...... ⎬	11s. 0d.	,,
Or to Bristol, ... ⎭	19s. 0d.	.,
Or to Truro, timber, ...	19s. 0d.	,,

PART V.

——:o:——

——:o:——

CHAPTER I.

FOREST LANDS.

A POPULAR name by which the *Pinus silvestris* is more extensively known than it is by the name of Scotch fir, is the Riga pine; but this latter designation is associated with tall straight trunks, such as are fitted for being used as masts and booms, while the former is associated with what, compared with these, seem dwarfed shrub-like trees unsuitable for such applications.

It is from the place of export that it has received that designation. Most of the timber exported thence, as has been mentioned, is brought thither from the provinces of Smolensk and Vitepsk, but some of it is the produce of the Baltic provinces of Russia. These are generally described as being three in number.

Conterminous with the government of Kovno, in Lithuania, situated between it and the Baltic, and the Gulf of Riga, is Courland. To the east of this, separated from it by the Dwina, lying between the north-western boundary of Vitepsk and the Gulf of Riga, is Livonia; and to the north of this, between this and the Gulf of Finland, is Estonia.

Limited and compressed as Estonia or Est-land, the land of the Ests, may now be, that tribe represents the primitive people inhabiting the whole of this region previous to the settlement within its bounds of the dominating German population.

Three, if not four, or more distinct tribes resident here are represented by these Ests, while a fourth allied family or tribe—the Fins—are found on the northern shores of this arm of the Baltic, inhabiting still the whole land between Russia and the Gulf of Bothnia, and giving their name to that territory. The Ests, or a large portion of them, and their territory constituted from the thirteenth to the fifteenth century a federal state, acknowledging the Roman Emperor and the Pope as its paramount lords and supreme heads. And the people differ as manifestly from the Lithuanian and Polish people of the adjacent provinces as they do from the Russians, whose empire extends over all.

Lithuania is frequently spoken of as a Polish colony, and I have no reason to doubt the correctness of the supposition that many of the more energetic Polish people penetrated into the country, contiguous as it is to their own, and in course of time became the dominant race. Something similar has occurred here—the dominant race is composed of the descendants of German emigrants, who, in turn, are being dominated by Russia.

Including the outlying islands, these provinces cover an area of 1,754 German miles, and comprise a population numbering about 1,850,000.

Courland is described as a flat country, interspersed with sand hills, extensive heaths and marshes, and patches of cultivated land, intersected by two ranges of hills running east and west. About two-thirds of the surface are covered with noble forests of pine and fir, interspersed occasionally with groves of oak. Game is abundant; and deer, bears, wolves, lynxes, and martins, haunt the forests. Agriculture and the rearing of cattle are the leading branches of industry.

According to official returns, which I obtained some years ago, in Courland the total area of forests is 851,000 desatins, the area of crown forests is 458,963 desatins, giving for every square verst 35·4 desatins of forests, and 19 desatins of crown forests—giving 1·4 desatins of the former, and ·7 desatins of the latter, for every inhabitant of the province. The annual fellings in the crown forests yield 61·8 cubic feet of wood, and 52 roubles 9 kopecs per desatin.

On crossing the Dwina, travelling eastward, we enter Livonia, a province, the physical aspect of which is different, and also the people, who manifest more of an Estonian character, while their language is designated Dorpat Estonian, in contradistinction to that spoken in Estonia, which is designated Revel Estonian.

According to official returns which I obtained at the same time with those already cited, the total area of forests in Livonia is 1,896,000 desatins ; the area of crown forests is 195,711 desatins ; giving for every square verst 47·4 desatins of the former, and 4·8 desatins of the latter, and giving 1·9 desatins of the former, and 0·2 desatins of the latter for every inhabitant of the province. The annual fellings in the crown forests yield 30·7 cubic feet of fuel ; and the revenue is 32 roubles 1 kopec per desatin.

By a boundary imperceptible to the passing stranger is Livonia separated from Estonia, the province situated further to the east, with Revel for its capital.

In Estonia the total area of forests is 450,000 desatins, the area of crown forests is 4,556 desatins, giving for every square verst of territory 26 4 desatins of forests and 0·2 desatins of crown forests, and giving 1·3 desatins of the former and 0·1 desatins of the latter for every inhabitant of the province. The annual fellings in the crown forests yield 44·2 cubic feet, and the revenue is 20 roubles 6 kopecs per desatin.

Both in Courland and Estonia are extensive forests of the lime tree, and these have given rise to much local industry, but, as has been intimated, there are also forests yielding both forest timber and firewood.

CHAPTER II.

FOREST ADMINISTRATION.

As has been stated, the administration of the management of forests in Russia constitutes a special department of the Ministry of Imperial domains. But there are in different parts of the Empire forests which were formerly, and still continue to be, under the supervision of special departments of the administration. Thus has it been with the government forests in the Baltic Provinces. In a codification of the then existing forest laws of the Empire, issued in 1857, it was enjoined in regard to the fourth division of the admiralty forests in the Baltic district :—

'336. The admiralty forests of the Baltic district are supervised by a board of management, similar to the boards of other districts.

337. The members of this board are determined by the authorities. The requisite number of forest officials and warders, for the improvement and protection of the woods, are appointed by the board of managers.

338. The Board is located in the town of Vytegra, as the most suitable for its duties ; but the Ministry of Imperial Domains may appoint any other town as the seat of the board.

339. The principal duties of the board are :—

1. Supervision of admiralty woods and forests, which will be placed under its sole control and management, in the governments of Olonetz, Novgorod, and St Petersburg.

2. Preparation of leaved and other timber in the government of Olonetz for the port of St Petersburg.

3. The allotment of civil forests, as well out of the district as out of the government forests, for the ports of Revel and St Petersburg.
4. Preparation of timber (*a*) in Courland and the island of Osel; (*b*) on the Dwina, until the appointment of the western board of management; (*c*) along the Nieman river, from the confluence of the same with the Windau canal; (*d*) from Finland, should such preparations be deemed desirable.

340. The following districts come under the exclusive control of the Direction of the Baltic district:—
1. All eastern and northern leaved and other forests, from which timber can be conveyed to St Petersburg, after the distribution of lands to peasants, namely, lands producing less remunerative qualities of timber.
2. Several wooded districts of St Petersburg government already apportioned to the Morin department.
3. Forest districts and parks which have to be set aside, on account of adaptability for timber being floated down by flowing rivers in Olonetz, Novgorod, and St Petersburg governments, after deduction of lands apportioned to government peasants, and for building of government and private barges, also for sawmills, and for Olonetz and St Petersburg government foundries.

341. Out of the remaining portions of forests of this district it is permitted to prepare timber for building of ships, according to requirements, but without exhausting the forests. The timber requisite for government building purposes at Revel is to be furnished as far as practicable from government forests.

342. As regards management and improvement of forests, the division of the same into forest districts, and

the preparation of building timber, the board of management will be guided in its action by—

1. Rules existing for other Admiralty forest districts.
2. Rules issued to the existing commission of Pudosk town, and pre-sketched in code, &c., of Department of goverment forests.
3. General forest laws.

The Department of admiralty forests is, however, directed to issue, without loss of time, a codex for the Baltic districts, with basis of rules existing for northern districts of leaved forests.'

In Courland there are several forests belonging to the government, and in the codification of the then existing laws in regard to administration and management of forests, the following were published as the laws and regulations relating to these forests, specifications being given of the chapter and section of the law on which each statement was based :—

' *Foundation of Board of Management of Government Forests in the Government of Courland.*

DIVISION I.—GENERAL COMPOSITION OF BOARD.

' 362. The local management of government forests of Courland belongs to that of the Courland Palata Imperial Domains, which manages these forests on strength of rules 867 to 1,050, 1,534 to 1,589, 1,645 to 1,668 of this codex.

363. In forest districts of Courland there exist, (1) foresters, according to rules of forest corps, as stated in section 70-71 ; (2) foresters and under foresters, with warders and guardians.

Note.—The foresters and foresters' assistants presently existing are to be replaced, on their leaving the service, by students of the forest corps

of the St Petersburg Institute, and these may
serve up to grade of Lieutenant-Colonel.

364. All the governments constituting Courland are divided
into twenty-nine forestries. To each is attached
a forester, two under foresters, according to size of
district, and two or three warders on horseback.
The foresters are appointed by the forest department of Imperial Domains on recommendation of
Palata; the rest are appointed and dismissed on
option of Palata.

365. Forest warders are appointed to watch over the conservation and security of government forests ; and
these are elected from holders of government
dairies or farms situated within the limits of the
district. The number of these to be fixed by
Palata of government domains, subject to approval
of local governor, in proportion as lands are
planned and mapped out, and the situation
becomes permanent ; and in order not to render
the situation complicated, they are to be excluded
from register of proprietors, for the time being, and
to be subject to the foresters, whose fields and
meadows they are required to till and cultivate.
In this office are to be placed only the best,
sober, and best behaved peasants, who are
irremovable from their position, and can only lose
their situation through bad behaviour and faults,
in which case their place is to be filled by others, by
arrangement of the Palata with the occupant for
the time being, and on the approval of the local
governor of the government.

The forest warders pay in the government
taxes to the forester, and the money due to the
owner for the time being they pay to the latter by
inventory—the so-called *wakengeld*—which money,
on the separation of the forest warders from the
dairies, is not to be deducted from the inventory ;
and the foresters pay over, or send those monies
to where they belong.

366. The exclusion of the forest warders from the lists of proprietors for the time being is to be effected at the close of the terms of rental, but should it be necessary to relieve certain warders from their duties towards the proprietors for the time being before the term, then the Palata of Imperial domains must, having solicited permission from the governor, notify the forest department of Imperial domains of the contemplated act, in order that the minister of these domains may, in reference to the government Palata, settle accounts with the holder of the dairy or farm, and reduce the rental in proportion.

Of Duties and Emoluments of Foresters and Under Foresters, and other Forest Servants.

367. Foresters, under foresters, and warders on horseback, have their houses situated on lands belonging to the government forests. Each forester (apparently superior officer) has 10 desatins of arable land allowed him, and meadows that yield 100 loads (or 30 poods) of hay. Under foresters are allowed one-half the allowance of a forester. The horse warders are allowed 3 desatins, and of hay 25 cart loads. Besides this, the foresters are allowed 5 per cent. of the gross income of their circuit, and in case of their exerting themselves, they may receive extra remuneration.

368. The Palata of Imperial domains is to assign a sufficient quantity of timber for the building and repair of foresters', under foresters', and warders' houses. The requisite supplies of workmen are furnished from the government farms, which receive building timber and firewood from the forest circuit. The interior or ornamental work of their house is to be done by the foresters, &c., but this alone at their own expense. The sum expended

is to be charged to the individual on the basis of a
twenty years' service, so that if a forester or under
forester, &c., leave the service, he will be charged
for the value of the house for as many years as he
has not served of the number of twenty years;
and the new comer is bound to pay his prede-
cessor, who ,has vacated by reason of death or
otherwise, for the remaining years up to the full
twenty years. For this reason the forester is
bound to give to the Palata of Imperial domains
an account of the sum expended in building his
house.

Note.—The foresters, &c., of the Baltic provinces,
are not subject to these rules, as houses, &c., are
furnished to them on other conditions.

369. The cultivation of the fields and hay fields of the
foresters and under foresters is the duty of the
forest warders, inasmuch as these warders seldom
cultivate more than one desatin for themselves,
but the work of ploughing, reaping, and har-
vesting cannot be executed by the warders alone.
Therefore the Palata of Imperial domains is
required to calculate, on the basis of the amount of
land and the number of forest warders, how much
land each forester and under forester has unculti-
vated, in order that the necessary workmen may
be sent from the government farms, which receive
assistance in timber and firewood. On the basis
of this calculation of ploughed land, hay-fields,
and corn lands, the Palata sues for and allots all
the work and working days between the government farms in proportion to the number of
peasants' houses.

370. The warders on horseback—mounted warders—are
bound to till their own fields.

371. The Palata of Imperial domains is diligently to see
that foresters and under foresters do not exact too
many duties according to schedule from the

forest warders, nor demand too much assistance from government farms, in order that all the labour may be done according to correct time.

372. The foresters and under foresters may send the forest warders with their produce to market; but not with more than six quarters of grain during the whole winter.

373. The forest warders are under the immediate and full control of the foresters and under foresters, and are bound to guard the forests, and to ride about them continually, and to execute their superior's commands without fail; and, as the foresters and under foresters are responsible for the neglect of the warders, they are at liberty to inflict slight fines on the latter, and in case of greater offences, to inflict corporal punishment, with permission of the Palata of Imperial domains, and in case of unfitness, to demand their expulsion from the service.

374. The Palata of Imperial domains requires the government land surveyors in preparing the plans, to suggest the most suitable track to be allotted to the forest servants, warders, and inspectors; these plans are laid before the Palata, who again refers them for sanction to the fourth department of Imperial domains. The proprietors for the time being of the government farms, or their managers, are to allow the forest inspectors as many workmen as are allowed to other government peasants : which allowance cannot be altered without the sanction of the Palata.

375. The forest inspectors wear brass badges on their coats that they may be recognised, and that no opposition may be offered to them in the execution of their duties.'

CHAPTER III.

OF Revel and Estonia, and of life there, a pleasing picture is given in a small volume published in 1844 under the title of *Letters from the Shores of the Baltic.* The quantity of forest land in this province is small in comparison with what there is in the more western provinces; but it is not altogether devoid of forests, and in general aspects at the different seasons of the year, a description of this may be accepted as more or less descriptive of the characteristics of all. The authoress of that work thus writes of a day spent in the country shortly after her arrival :—

' Summer's busy workshop has long been closed, and Nature has shrouded herself deep within her monumental garments, though, as if to cheer us on the long and dreary winter pilgrimage before us, she charitably reveals a few glimpses of her real features, shows us here a line of bold grey rocks butting through the snow, and there a dashing cascade, which the frost has not yet completely stiffened, till our faith in her hidden beauties is only equalled by our impatience to behold them.

' There is something, however, very exhilarating in this breathless, still, bright cold—with a clean white expanse —a spotless world before you—every tree fringed—every stream stopped—freedom to range over every summer impediment ; while the crystal snow, lighting up into a delicate pink or pearly hue, or glistening with the brightest prismatic colours beneath the clear, low sun, and assuming a beautiful lilac or blue where our long shadows intercept its rays, can no longer be stigmatised

as a dead lifeless white. We walk every day, and no
sooner are the heavy double doors, which effectually seal
our house, heard to open, than half a dozen huge, deep-
mouthed cattle-hounds come bounding through the deep
snow to meet us, oversetting, with the first unwieldy caress,
some little one of our party, scarce so tall as themselves,
and even besetting the biggest with a battery of heavy
demonstrations, to which it is difficult to present a firm
front. Sometimes we take the beaten track of the road,
where peasants with rough carts, generally put together
with less iron than an English labourer would wear in his
shoes, pass on in files of nine or ten ; as often as not the
sheepish-looking driver, with his elf locks, superadding his
own weight to the already overladen little horse—or where
a nimble-footed peasant woman, with high cap and clean
sheepskin coat, plunges half-leg high into the deep snow
to give you room, and nodding, and showing her white
teeth, cheerily ejaculates *Terre hommikust*, or Good day.
Or we follow a track into the woods so narrow that we
walk in each other's steps like wild Indians, and the great
dogs sink up to their bodies in the snow whilst
endeavouring to pass us. This is the land of pines—lofty,
erect battalions—their bark as smooth as the mast of a
ship—their branches regular as a ladder, varying scarce an
inch in girth in fifty feet of growth—for miles interrupted
only by a leaning, never a crooked tree—with an army of
sturdy Lilliputians clustering round their bases—fifty
heads starting up where one yard of light is admitted.
What becomes of all the pruning, and trimming, and training—the days of precious labour spent on our own woods?
Nature here does all this, and immeasurably better, for
her volunteers, who stand closer, grow faster, and soar
higher, than the carefully planted and transplanted children
of our soil. Here and there a bare jagged trunk, and a
carpet of fresh-hewn boughs beneath, show where some
peasant-urchin has indulged in sport which with us would
be amenable to the laws—viz., mounted one of these
grenadiers of the forest, hewing off every successive bough

beneath him, till, perched at a giddy height aloft, he clings to a tapering point which his hand may grasp. The higher he goes the greater the feat, and the greater the risk to his vagabond neck in descending the noble and mutilated trunk. In perambulating these woods, the idea would sometimes cross us that the wolves, the print of whose footsteps, intercepted by the dotted track of the hare and slenderly defined claws of numerous birds, are seen in different directions, and even beneath the windows of our house, might prowl by day as well as by night. One day, when, fortunately perhaps, unescorted by the huge dogs, we were mounting a hill to a neighbouring mill, my companion suddenly halted, and lying her hand on mine silently pointed to a moving object within fifty yards of us. It was a great brute of a wolf stalking leisurely along—its high bristly back set up—its head prowling down—who took no notice of us, but slowly pursued the same path into the wood which he had quitted a few minutes before. We must both plead guilty to blanched cheeks, but beyond this to no signs of cowardice ; and, in truth, the instances are so rare of their attacking human beings, even the most defenceless children, that we had no cause for fear. They war not on man, unless under excessive pressure of hunger, or when, as in the case of a butcher, his clothes are impregnated with the smell of fresh blood. This is so certain an attraction that peasants carrying butchers' meat are followed by wolves, and often obliged to compound for their own safety by flinging the dangerous commodity amongst them ; or if in a sledge, three or four of these ravenous animals will spring upon the basket of meat and tear it open before their eyes. Wherever an animal falls, there, though to all appearance no cover nor sign of a wolf be visible for miles round, several will be found congregated in half an hour's time. Such is their horrid thirst for blood, that a wounded wolf knows that only by the stictest concealment can he escape being torn in pieces by his companions. As for the dogs, it is heart-rending to think of the numbers which pay for their

fidelity with their lives. If a couple of wolves prowl
round a house, or fold, at night, a dozen dogs, with every
variety of tone, from the sharp yap of the shepherd's
terrier to the hoarse bay of the cattle-hound, will plunge
after them, and put them to flight. But if one, more
zealous, venture beyond his companions, the cunning
brutes face about, seize him, and before three minutes are
over there is nothing left of poor *Carrier Pois,* or sheep-
boy,—a common name for these great mastiffs,—but a few
tufts of bloody hair. The cattle defend themselves
valiantly, and the horses, and the mares especially who
have a foal at their side, put themselves in an attitude of
defence, and parry off the enemy with their fore-feet—their
iron hoofs often taking great effect. But woe be to them
if the wolf, breaking through the shower of blows, spring
at the throat, or, stealing behind his prey, fasten on the
flank !—once down all is over, though there be but one
wolf. Sometimes, in a sudden wheel round, the wolf will
seize upon a cow's tail, on which he hangs with his jaws
of ten-horse power, while the poor animal drags him
round and round the field, and finally leaves the
unfortunate member in his grasp, too happy to escape
with a stump. At one time these animals increased so
frightfully in number that the Ritterschaft, or assembly
of knights, by which name the internal senate of this
province is designated, appointed a reward of five roubles
for every pair of ears brought to the magistrate of the
district. This worked some change, and, in proportion as
the wolves have fallen off, the Ritterschaft has dropped its
price, though an opposite policy would perhaps have been
more politic, and now a pair of ears, generally secured
from the destruction of a nest of young ones, does not
fetch more than a silver rouble, or three roubles and a
half.* An old plan to attract them was to tie a pig in a
sack, squeaking of course, upon a cart, and drive him
rapidly through a wood or morass. Any cry of an animal

---

* Paper. Nevertheless a thousand wolves upon the average are killed in a year.

is a gathering sound for the wolf, but the voice of man, made in his Creator's image, will hold him aloof. The blast of a horn greatly annoys them, a fiddle makes them fly, and the gingling of bells is. also a means of scaring them, which, besides the expedience of proclaiming your approach in dark nights on these noiseless sledge-roads, is one reason why all winter equipages are fitted up with bells.'

Nor is the breaking up of the winter in Estonia less truly portrayed :—' The soft hand of spring imperceptibly withdraws the bolts and bars of winter, while the earth, like a drowsy child, 'twixt sleeping and waking, flings off one wrapper after another and opes its heavy lids in showers of sweet rivulets. And the snow disappears, and the brown earth peeps almost dry from beneath ; and you wonder where all the mountains of moisture are gone. But wait—the rivers are still locked, and though a strong current is pouring on their surface, yet, from the high bridge, the green ice is still seen deep below, firm as a rock—and dogs go splashing over in the old track, and peasants with their horses venture long after it seems prudent. At length a sound like distant thunder, or the crashing of a forest, meets your ear, and the words " *Der Eisgang, der Eisgang !* " pass from mouth to mouth, and those who would witness this northern scene hurry out to the old stone bridge, and are obliged to take a circuitous route, for the waters have risen ankle deep—and then another crash, and you double your pace regardless of wet feet, and are startled at the change which a few hours have produced. On the one side, close besetting the bridge, and high up the banks, lies a field of ice lifting the waters before it and spreading them over the country ; while huge masses flounder and swing against one another with loud reports, and heave up their green transparent edges, full six feet thick, with a majestic motion ; and all these press heavily upon the bridge, which trembles at every stroke, and stands like a living thing labouring and gasping for

T

breath through the small apertures of the almost chocked
arches.   On the other side the river is free of ice, and a
furious stream, as if all the imprisoned waters of Russia
were let loose, is dashing down, bearing with it some huge
leviathan of semi-transparent crystal, and curdling its
waters about it, till this again is stopped by another field
of ice lower down.

'The waters were rising every minute—night was
approaching, and the beautiful old bridge gave us great
alarm, when a party of peasants, fresh from their supper
at the *Hof*, and cheered with brandy, arrived to relieve it.
Each was armed with a long pole with an iron point, and
flying down the piles and on to the ice itself, began
hacking at the sides of the foremost monster, till,
impelled by the current beneath, it could fit and grind
itself through the bridge and gallop down to thunder
against its comrades below.   The men were utterly
fearless, giving a keen sense of adventure to their dangerous task which rivetted us to the spot ; some of the most
daring standing and leaning with their whole weight over
the bed of the torrent upon the very mass they were
hewing off, till the slow swing which preceded the final
plunge made them fly to the piles for safety.   Some
fragments were doubly hard with imbedded stones and
pieces of timber, and no sooner was one enemy despatched
than another succeeded ; and although bodies of men
continued relieving each other all night, the bridge
sustained such damage as could not be repaired.   All was
over in twelve hours, but meanwhile, " the waters prevailed exceedingly upon the earth," and every hill and
building stood insulated.

'Such was the picture of our life a fortnight ago, since
when a still more striking change, if possible, has come
over the face of things.   The earth, which so lately
emerged from her winter garb, is now clad in the liveliest
livery ; while every tree and shrub have hastily changed
their dresses in Nature's vast green-room, and stand all

ready for the summer's short act. Nowhere is Nature's
hocus-pocus carried on so wonderfully—nowhere her
scene-shifting so inconceivably rapid. You may literally
see her movements. I have watched the birdcherry at
my window. Two days ago, and it was still the same
dried up spectre, whose every form, during the long winter,
the vacant eye had studiously examined while the thoughts
were far distant—yesterday, like the painter's Daphne, it
it was sprouting out at every finger, and to-day it has
shaken out its whole complement of leaves, and is throwing
a verdant twilight over my darkened room. The whole
air is full of the soft stirring sounds of the swollen buds
snapping and cracking into life, and impregnated with
the perfume of the fresh, oily leaves. The waters are full
and clear—the skies blue and serene—night and day are
fast blending into one continuous stream of soft light, and
this our new existence is one perpetual feast. Oh, winter !
where is thy victory? The resurrection of spring speaks
volumes.'

The same pencil has painted the passing away of
summer:—' The beauties of autumn, and the moral of its
yellow leaves, are seen and felt in all countries. Nowhere,
however, I am inclined to think, can the former be so
resplendent, or the latter so touching, as in the land where
I am still a sojourner. In our temperate isle autumn
may be contemplated as the glorious passing away of the
well-matured—the radiant death-bed of the ripe in years
—while here the brilliant colours on earth and sky are
like the hectic cheek and kindling eye of some beautiful
being whose too hasty development has been but the
presage of a premature decay. Thus it is that the vast
plains and woods of Estonia are now displaying the most
gorgeous colours of their palette, ere the white brush of
winter sweep their beauties from sight, while the golden
and crimson wreaths of deciduous trees, peeping from
amongst the forests of sober pines, may be compared to
gay lichens sprinkling their hues over a cold grey rock, or

to a transient smile passing over the habitual brow of care.

'But all too hasty is the progress of this splendid funeral march—even now its pomp is hidden by gloomy slanting rains, its last tones lost in the howl of angry blasts, which, as if impatient to assume their empire, are rudely stripping off and trampling down every vestige of summer's short-lived festival, while Nature, shorn of her wealth, holds out here and there a streamer of bright colours, like a bankrupt still eager to flaunt in the finery of better days.'

FINIS.